EFFECTIVE STRATEGIES IN THE TEACHING OF MATHEMATICS

A Light from Mathematics to Technology

Velta Clarke

University Press of America,® Inc.
Lanham · Boulder · New York · Toronto · Oxford

University Press of America,® Inc.
4501 Forbes Boulevard
Suite 200
Lanham, Maryland 20706
UPA Acquisitions Department (301) 459-3366

PO Box 317
Oxford
OX2 9RU, UK

Printed in the United States of America
British Library Cataloging in Publication Information Available

ISBN 0-7618-2601-7 (clothbound : alk. ppr.)
ISBN 0-7618-2602-5 (paperback : alk. ppr.)

∞™ The paper used in this publication meets the minimum requirements of American National Standard for Information Sciences—Permanence of Paper for Printed Library Materials, ANSI Z39.48—1984

This work is dedicated to my immediate family and to all my past students who are now effective teachers of mathematics.

Contents

Part 1: Historical Context of Mathematics

Figures

Tables

Preface

Effective Strategies in the Teaching of Mathematics is an explication of all the knowledge and skills that middle and high school teachers of mathematics should possess. Teachers are not born, they are "created," and they must learn the many skills that constitute the repertory and the mystery of successful teachers. The enigmatic quality of good teachers may be acquired through the instructional process, and this book provides access to the key that teachers may use to become practitioners of effective instruction in mathematics.

The work is designed as a methodology course for preservice teachers who are seeking certification for teaching mathematics in middle and high schools. It describes the major philosophies, learning theories, and principles that are fundamental to teaching and explicates with examples the current methodologies of delivering instruction in mathematics. The book is based on the professional experiences of the writer, who has taught methods of teaching mathematics with technology to preservice teachers and has supervised student teachers in the field for several years. The author has developed an appreciation for the methods that are required for effective teaching of mathematics and has enshrined them within the pages of this book.

The main thesis of the work may be described as using "empiricism or the spirit of utility" as fundamental to mathematics pedagogy. It proposes that mathematics had its genesis in empirical preoccupations of early civilizations and was invented to solve their practical problems. As the center of civilization moved from the Middle East to Europe, mathematics transformed its empirical base and assumed more theoretical and sophisticated proportions: it became a science, drawing inspiration from within itself and using methods of research, axioms, and rigorous proofs. Mathematics grew into a cultural area of university study and entered the realm of academicians, becoming less accessible to the ordinary learner. Generally students fear mathematics and see it as alien to their lives. The perception that mathematics only belongs to the cultural realm, however, is a fallacy. In fact, mathematics pervades every aspect of the real world where it is used to solve most economic and social problems. Notwithstanding the critical importance of mathematics as an instrument for finding answers to the recurring sources of social perplexity, mathematics is usable mainly by members of cultural elite who have been exposed to its mysteries in the education process. For most others, mathematics is a fearsome subject to be avoided.

The key to the epistemological aspect of mathematics is in the pedagogy. Knowledge of mathematics content is important, but for students to acquire this knowledge, it must be taught effectively by good teachers. In the current society, when there is great emphasis on "reality" as an instructional strategy, the belief is that mathematics should be taught as an empirical subject so as to make it meaningful and available to every citizen as a means of serving practical concerns and also enriching their lives.

There is a general debate among educators and professional organizations such as the National Council of Teacher of Mathematics (NCTM) concerning effective methods for teaching mathematics. The NCTM has developed standards that include mathematics as an empirical subject. The report of the Third International Study of Science described the quality methods used by Japanese teachers who emphasized understanding mathematics and employed strategies of problem-solving, and real life experiences. These educators and organizations have supported using empirical methods of mathematics instruction. This work contributes to the debate about mathematics instruction and proposes viable suggestions for effective mathematics pedagogy.

The process of making mathematics available to everyone should have its genesis in schools. It should begin as the responsibility of the teacher of mathematics who must produce impressive achievements by students. Pedagogy should return mathematics to its empirical base, and teachers should impart the subject by relating it to the social context as well as to the experiential realities of all students. An empirical application of mathematics, involving concrete representation, constructivism, concept attainment, discovery, experimentation, experiential backgrounds, induction, inquiry, problem-solving, and use of social realities as a context, is seen as critical to the learning of mathematics as an instrument for the management of everyday situations and transactions.

It is generally agreed that the "time-honored" didactic methods have conceived of mathematics as a cultural subject. These methods, which are generally called "Chalk and Talk," involved providing students with a definition followed by explanations and examples by the teacher, while the students, if they could, simply absorbed, rather than created within their minds, the new knowledge. These methodologies were based on traditional philosophies and learning theories, and although they always serve a purpose, they should give way to modern methods that are based on more pragmatic philosophies, learning theories, and theories of how the brain functions. In addition, studies and research in cognitive science and neuroscience have provided new perspectives on how individuals learn and have spawned new models of learning based on information processing.

This work explores new and dynamic empirical learning and methodologies and the application of technology to learning, and exemplifies them as viable pedagogy in this work. *Effective Strategies in the Teaching of Mathematics* should be essential reading for all teacher candidates and should occupy an honored place in the library of all mathematics professionals as well as progressive educators.

Acknowledgments

This book was the outcome of interaction with my many students of mathematics education, both in the classroom and during student teaching in the field. They demonstrated how the methods advocated in this book may be used. The ideas are not merely theoretical; they may be implemented by resourceful and enlightened teachers. Classroom teachers with whom I have interacted and who acted as cooperating teachers for my students also made comments on the methods and influenced my ideas. Many of them described the ever-changing standards in mathematics and the new assessment methods.

I am indebted to my colleague Dr. Kathleen Velsor who read the manuscript and gave critical review. I owe a debt of gratitude as well to Dr. Annette Mahoney for her careful reading of the manuscript. Mathematics teachers Daniel Bloomfield and Marion Johnson read the sections on mathematics pedagogy. I would also like to thank Marcy Thorner, The Grammar Guru, for her dedication to perfection in the editing and formatting of the book.

Introduction

It is a generally acknowledged truth that a good teacher is the product of the educational process. A teacher candidate is conditioned by the many philosophies, theories, and practices learned in the training process. *Effective Strategies in the Teaching of Mathematics* provides a comprehensive set of learning experiences for all those teacher education candidates who aspire to be good teachers of mathematics.

This book describes the systematic development of two major themes that are inextricably intertwined and concurrently applied to mathematics pedagogy. The first is the historical growth of mathematics, from the nascent beginnings using empirical methods that characterized early civilizations, to more formalistic symbolic methods applied in Europe. Modern mathematics and technology are a direct outgrowth of the development of mathematics. The second theme is the development of learning theories. Philosophy supplied the earliest explanation of learning. With the emergence of science, philosophy gave birth to psychology and to behaviorist and cognitive learning theories. Modern methodologies emphasize the role of the brain in learning through such methodologies as neuroscience and information processing. *Effective Strategies in the Teaching of Mathematics* is an explication of how these two themes merged and how they may be applied in the teaching learning process.

The main ideas in this book are derived from philosophy, psychology and science. From them are derived the principles and methods of teaching. The structure of the lesson plans, however, are the original idea of the author and the contribution to knowledge.

The work contains four developmental sections. The first section is historical and gives an account of the beginnings of mathematics in antiquity and traces its growth through several civilizations. It describes how empirical mathematics began and show how it was transformed by the successive civilizations.

The Babylonians and Egyptians initially practiced numeration, computation, and mensuration, which grew into algebra and geometry; observation of the heavens led to trigonometry. Other empirical forms of mathematics were concrete representations of the Chinese, linguistic and visual representation of the Indians, the deductive logic of the Greeks, and architecture and military preoccupations of the Romans. It was Islam that synthesized the concrete of the Chinese, the linguistic theories of the Indians, and the logic of the Greek and transformed them into a new form usable by the early modern Europeans. Empirical mathematics gave way to rational and axiomatic and symbolic mathematics of the modern age. Mathematics also gave rise to technology as a new method for solving problems and today is the chief means for finding answers to difficult economic and social realities. In fact, it is generally believed that mathematics will become almost anachronistic and will be replaced by technology as the main economic instrument. Those who know mathematics and its relationship to technology, however, will be the major beneficiaries as well as contributors to the new world order.

The second section explicates the foundations of pedagogy. It describes philosophy as the basis of pedagogy. Psychology was born when science touched philosophy, and from the new synthesis there emerged many theories of learning based on scientific principles. Science also provided a basis for learning theory with the study of the brain and how man learns.

The third section is based on pedagogy. Its concern is mostly practical. The section describes the principles of learning that grew out of the philosophical and psychological theories and the functioning of the brain. This is the section that describes the skills that may be implemented in schools. It describes how to make unit and lesson plans and gives examples of these, describes curriculum models, how to make tests, and methods of alternative assessment and reflective teaching. It describes the historical development of the school curriculum, methods of discipline, and how to create a dynamic learning environment. The last chapter describes how the teacher may conduct experimental research as a means of solving problems in the classroom.

The final section describes the development of technology and gives detailed examples of how to implement it. It provides instructions for the novice on how to use Microsoft Excel and gives examples of it that may be used in the solution of mathematical problems such as quadratic and simultaneous equations.

The preservice teacher will find this a work a learning experience and a practical book that may prepare the novice for teaching effectively in middle and high schools. Diligent pursuit and application of the methods explicated in the work will no doubt help beginning teachers to become thoughtful and grow into being effective and consummate professionals.

Part 1

Historical Context of Mathematics

Chapter 1

Empirical Context of Mathematics

Mathematics of the modern age is a legacy of the great civilizations of the past. Fundamental concepts were developed independently in several countries, others were transmitted through acculturation, and still others represented the outstanding contributions of a single civilization or age. Nevertheless, methods of calculations and arguments show remarkable similarity from one culture to the next. The timeline had its genesis with the ancient Babylonians and Egyptians in the fourth millennium B.C.E. In the fifth century B.C.E., mathematics flourished simultaneously in three civilizations: Greek, Chinese, and Indian. Greece continued until the fourth century C.E. and then gave its legacy to the Romans. The Chinese and the Indians continued until the thirteenth and sixteenth centuries respectively.

Western civilization made its greatest contribution beginning with the Greeks, and after centuries of quiescence reclaimed hegemony with the onset of the Renaissance of the fifteenth century (Struik, 1987). Mathematics was a practical response to the prevailing needs and preoccupations of each civilization in each historical epoch ranging from the empirical or "spirit of utility" of the Babylonians and Egyptians to the "rational" of the Greeks to the "axiomatic" of this modern technological age.

Theory

This work proposes that mathematics is both a cultural and utilitarian subject. Over the centuries it transformed its preoccupations from practical to theoretical and in the process has itself become a science. It is proposed that the best way to appreciate a subject is to examine its fundamental ideas from their nascent stages and to trace the historical development of these ideas over time so as to rediscover the context and situations in which they were formulated and flourished. In describing the progressive growth of mathematics, the pivotal central concern will be *why* the various principles evolved and the nature of the practical problems to which they were applied in each historical period. The outcome of this discussion is to explicate the theory that in each historical period, mathematics has been related to the real-life preoccupations of the inventors. In this regard, the current epistemology should be consistent with the development of these very critical concepts and methodologies.

Historical Origins of Mathematics

Mathematics had its origins in the practical preoccupations of human activities of ancient civilizations. As the old agricultural villages became anachronistic and cities grew, there emerged the need for more sophisticated forms of economic, educational, political, and social organizations. Trading communities and established institutions gave motive force to the highest imaginations and the invention of dynamic and creative instruments of operating, including mathematics, which became necessary to control the burgeoning environment. Mathematics was a response to man's need to keep track of the ever-increasing complexity of the transactions of developing civilizations, and it became the handmaid of administrators, engineers, and merchants. Earliest mathematics was based on arithmetic and included numeration, computation, and mensuration. As arithmetic became more complex, algebra emerged as a developmental stage of arithmetic as the response to solving equations describing abstract and sophisticated data. Mensuration gave birth to geometry, and the practice of observing the heavenly bodies and methods of controlling the floods grew into astronomy and trigonometry.

Algebra progressed through three developmental stages. The first stage, which was invented by the Babylonians and Egyptians, was based on the rhetorical or verbal style. The second stage, which was called the syncopated style and involved abbreviations, was introduced by the Greek Diophantus of Egypt in 250 C.E. It was also used by Indian mathematicians such as Bramagupta.

Modern symbolic algebra was the third stage and was introduced in Europe in the sixteenth century by innovators such as Cardona, Robert Recorde, and Christian Rudolf.

The developmental timelines overlapped across civilizations. Earliest mathematics began in Babylonia and Egypt between 3,500 and 1,000 B.C.E. Mathematics existed contemporaneously in China and India from the fifth century B.C.E. to the thirteenth century C.E., and in the Greek and Roman Empires from the fourth century B.C.E. to the first century C.E. The theater then shifted to the Islamic Empires of the Middle East between 700 and 1200 C.E., and finally to Western Europe after the decisive defeat of the Moors in Spain on 16 July 1212 C.E. With each change of scene, there was a shift in the method of computation and the style of presentation of concepts and principles, but across civilizations, the product was consistently the application of mathematics.

Utilitarian Mathematics

Mathematics had its earliest origins in Middle Eastern civilizations along the Tigris-Euphrates and Nile River valleys in Babylonia and Egypt respectively, during the fourth millennium B.C.E., where it was the very lifeblood of the existence of these civilizations. Its invention was precipitated by practical necessity of solving problems in everyday pursuits. In agriculture there were matters of controlling floods; drainage of swamps; and processing, storage and distribution of harvests. In engineering there were problems of building dams, irrigation, and construction of sewage systems and reservoirs. Administrative preoccupations included the coordination of such activities as collection of taxes and public works, the development of a calendar, and studies in the heavenly bodies. There was also specialization of functions in commercial activities and business transaction such as counting cattle, exchanging of money and keeping of accounts of the seasons (Struik, 1987, p. 212). Success in agriculture was dependent upon the study of heavenly bodies.

Mathematics in Babylonia

Babylonia, occupying the area of the fertile crescent between the Tigris and Euphrates rivers was the earliest known civilization. Around the Euphrates in the southern region known as Mesopotamia, there emerged several dynasties including the Sumerians between 4,500-2370 B.C., and the Akkadians from 2370-2200 BC, based on the city of Ur. They were conquered by the Amorites, also called the Babylonians, who founded the city of Babylon in 1990 B.C. Their greatest king was Hammurabi who reigned from 1792-1750 BC. In 1595 B.C., the region was invaded by Indo-European Kassites who maintained control over southern Mesopotamia for four hundred years, and Babylonia and was not again an important political power until the seventeenth century. Today, ancient Babylonia is known as Iraq.

In the upper Tigris area, north of Mesopotamia, based on the city of Nineveh, there emerged the civilization of the Assyrians in 750–613 B.C.E. In 587 B.C.E., they were conquered first by the Chaldeans and then by King Nebuchadnezzar of Babylon in Mesopotamia. Beginning in 549 B.C.E., there emerged in this region north of Mesopotamia, a new kingdom of Persia. Through the efforts of its military leader Cyrus, Persia created an extensive Empire. In 539, Cyrus conquered the Chaldeans and added Mesopotamia to his empire. This was followed by the conquest of Syria, Palestine, and Phoenicia; and in 525 B.C.E., and his son conquered Egypt. The Persian Empire existed from in 549–333 B.C.E., when it was conquered by Alexander the Great of Greece in 490 B.C.E. and continued under Hellenistic influence until it was destroyed by the Arabs in 637 C.E. Today, the land of Persia is known as Iran.

The ancient Babylonians in Mesopotamia used cuneiform writing, which is a curious form of wedge-shaped script etched on clay tablets. The mathematician Neugebaur, who lived between 1899 and 1992 C.E., developed a complex text in cuneiform . It was in Babylonia that a very sophisticated form of mathematics developed. Examples include

- Decimal system with symbols based on a positional system, for example, 1 = unit; < = 10

- There was no symbol for zero. The positional system was base 60. It was a sexagesimal system that gave birth to concepts of the 60-minute hour and 360 degrees in a circle.
- Fractions were written in sexagesimals.
- Square roots were in sexagesimal fractions.
- The binary system and how to recover numbers that had been combined, which led to quadratic equations.
- Tables were used in computation, for example multiplication tables, reciprocals, cube roots, and the pi.
- There was knowledge of how to find compound interest on annuities, for example, how long would a sum of money double itself at 5 periods of time?

It was in Babylonia that rhetorical style mathematics that involved the practice of indicating concepts in descriptive forms, originated. For example, "The square of the unknown quantity is the sum of 6 and 10." Based on cuneiform writings found on clay tablets during the reign of King Hammurabi (1700 B.C.E.), algebra concepts were described in parallel forms with words on the left-hand side and mathematical computation on the right (Baumgart, 1969, p. 243). The rhetorical method gave the problem, described the data in words, and the explained the methodology using numbers, the stated and check the solution. The following example was taken from clay tablets written during the reign of King Hammurabi has by Baumgart (1993, p. 235) in translation:

- Given length and width, I have multiplied length and width, thus obtaining area 252. I have added length and width to obtain the sum of 32. Required are the length and width.
- Given 32 sum, and ($X+Y=32$), 232 the area ($XY=232$)
- Answer 18 lengths, 14 widths
- Check: $18 \times 14 = 252$

An explanation of the method in parallel form then followed. Babylonians developed great skill in the solution of systems of equations with two unknowns, usually X and Y, using the parametric method. Their work demonstrated knowledge of square roots, cube roots, reciprocals, and exponential functions. It was from them that we derived notions of reckoning time into sixty minutes and seconds.

Egyptian Mathematics

Most of what is known of earliest Egyptian mathematics is derived from two historic scrolls written on papyrus: the Rhind or Ahmose Papyrus and the Moscow Papyrus. The Rhind is a scroll about twenty-eight feet by thirteen feet, written in hieratic script by Ahmose (1570–1546 B.C.E.), with 85 problems. It was acquired by Alexander Rhind, a British Egyptologist at Luxor, Egypt in 1857. One section of the scroll was transferred to the British Museum and another is in the New York Historical Museum. The Moscow scroll, which was first the property of Goldenischew, is now in the Moscow Museum. The scroll, which predated the Rhind, was written around 1850 B.C.E., and has twenty-five problems. The Rhind explains the principles of Egyptian mathematics thus,

- The addition principle.
- Fractions were computed. The fraction was a unit of 1, and there was a table of the calculated fractions.
- The decimal system was based on letters of the alphabet:
  V = 5, III = 10, C = 100, M=1000.
- Multiplication was based on repeated addition for example,
  1 28
  2 56
  4 112
  7 196
  In effect 28 * 7 = 196
- Conversely, multiplication of fractions was by repeated subtraction.
- Arithmetical multiplication was based on additions, for example,
  2 + 2 + 2 = 6 means 3 * 2 = 6

Like the Babylonians, Egyptians knew how to solve equations with two unknowns that they called "hau" and "aha." They made spectacular inventions; for example, they discovered that 360 degrees make a circle, and they devised methods of computing areas, triangles, cylinders, and volume. Egyptians demonstrated in their construction of the pyramids, the formula for the frustum, where "a and b are the lengths of the sides of the square and h is the height." This is described in the Moscow scroll as

$$V = \frac{h}{3}(a^2 + ab + b^2)$$

The Egyptians, too, used rhetorical or verbal style in algebra in the early stage. It was also in Egypt that the second stage of algebraic development occurred. It was invented by Diophantus, a Greek, who lived and studied in Alexandria. He pioneered the method of presenting algebra by abbreviations called syncopations as early as 250 C.E. The method is parametric and involves stating all unknowns from which an equation is derived. For example, he wrote, "'AY' to mean X^2 and "M" means constant term."

He also used the special symbols "plus of minus" to denote the square root of -1 and he also used "minus of minus" for its negative. Cooke (1993, p. 313) disputed that it was Diophantus, not the Arabs who invented Algebra.

The Chinese and Concrete Representation

Chinese civilization developed in isolation and was not affected by western influences until the Tang Dynasty in the seventh and eighth centuries C.E. Though independently developed, Chinese mathematics was remarkably similar to that of other civilizations. The history of China extends 6,000 years into the past. As with the civilizations of Mesopotamia and Egypt, the Chinese geography posed challenges and presented critical opportunities from which mathematics was invented. Like them too, mathematics was derived from practical considerations such as surveying, commerce, and administration. Similar to the Greeks, practical mathematics gave rise to sophisticated theoretical forms.

Earliest mathematics began in the dynasty when the tortoise was romanticized. It was a diagram called the *Lou Shu*, given to the Emperor to be used in surveying project for the control of floods. It was the magic square with numbers from 1 to 9 for rows and columns. The Emperor used the gnomon (square) and a compass in surveying. There is archaeological evidence that during the Shang Dynasty during the sixth to eighth centuries B.C.E., Chinese had in place a decimal system with base 60, also a counting rod, and mechanical devices for computation.

Their mathematics was based on concrete representations of practical squares, circles, and they invented what is today called matrix models. For example, trapezium was represented by a "dustpan-shaped field," frustum by a "square field," and a tetrahedron by a "turtle shoulder joint" (Eves, 1989). From these basic machinations, the

Chinese constructed formulas. Based on their penchant for concrete representations, they expressed numbers by using a bamboo stick placed on a board with an empty space for the zero.

This system was later transmitted to Europe. They represented numbers by transformation. They were interested in computation and verifying results, which, though not deductive, was in effect a kind of proof. The Chinese lattice design, for instance, indicated the presence of shapes such as triangles, hexagons, and parallelograms, without reference to deductive system. Nelson (1995) quoted a Chinese mathematician Chu Shib-chieb who used iconic forms such as a pile of reeds to represent a triangle with base N reeds and height N + 1 reeds. A second similar pile placed against this formed a rectangle. Thus,

> A triangle with base N and side N + 1
> A second triangle with base N and side N + 1 were placed side by side
> A rectangle is formed.
> The area would be N (N + 1) reeds.

By the third century B.C.E., during the Han dynasty (206–220 B.C.E.), Chinese were engaged in geometry. *The Arithmetical Classic of the Circular Path of Heaven* (Zhoubi Suanjing) is the earliest document of Chinese mathematics and was based on geometrical applications to astronomy especially with the use of the sundial or gnomon. The text states that the Emperor Yu used the gnomon for controlling the flood. Based on a *Book of Crafts*, names were given to angles. For example the right angle was *Ju* and half the right angle was *Xuan*. Other angles were obtained by increasing Xuan by fifty percent in three stages: they were $67^0\ 30^1$ called *zhu*, $101^0\ 15^1$ called *ke*, and $151^0\ 52^1$ called *quinzhe*.

The best known Chinese text assumed to be compiled in the first century C.E., during the Han Dynasty, is the *Nine Chapters on the Mathematical Art* (Jiuzhang Suanshu). Liu Hui dates the book in the third century B.C.E., contemporaneous with the writing of the Egyptian Ahmose Papyrus. Like the Ahmose, the book consists of practical and theoretical problems that were described and solved. The first chapter deals with field measurement (*fang tin*) and shows how to compute area, circles, spheres, and deals with the use of pi in computation. The other eight books deal with "cereals," "distribution of property," "width," "construction," "fair tax," "excess and deficiency," and "rectangular arrays."

The Nine Chapters also deals with algebra. There are 246 problems on algebraic topics such as linear equations, square roots, and quadratic equations. It includes topics on sphere and volume that are similar to the method used by the Greek Archimedes. Liu Hui added another chapter to *The Nine Chapters* called the "Master Suin Document" (Sunzi Suanjing) in which he gave problems of linear equation and treated surveying and the Remainder Theorem, which is a problem in which a number is divided by 3 with remainder 2; and divided by 455 with remainder 3, and divided by 7 with remainder 2. The Chinese used this system to solve simultaneous equations with two unknowns.

The Chinese used the *tangram* to demonstrate the Pythagorean Theorem that was also known to the Babylonians and was deductively demonstrated by Pythagoras himself. They also knew the triangle that has been attributed to Pascal. The figure that appeared in a Chinese book on mathematics was invented by an eleventh century mathematician named Yang Hui. The famous triangle, the basis of the binomial theorem, was commonplace in China.

The Chinese reached a high level of sophistication in the astronomy. This applied branch of mathematics was solely controlled by government, and today Chinese have the most extensive astronomical records in the world. Astronomy was based on the sun and the establishment of a lunar calendar. From the real-life problems Chinese also created quadratic functions and instruments for measuring were that were widely used during the Yuan dynasty in the thirteenth century C.E.

By the end of the thirteenth century the Greek Empire declined, but civilization continued to flourish in China. The Mongols, who were Muslims, conquered China and brought to that country contact with the intellectual achievement of the Muslim world. Knowledge was interchanged. Contact with the west began when the Jesuits arrived in China in 1582.

India and Linguistic and Visual Representations

Much of what is known of Indian mathematics is of recent origin, that is, beginning in the first millennium B.C.E. Undoubtedly, mathematics had existed for a long time, but little was preserved because the Indians, like the Chinese, wrote their mathematics on perishable materials such as the bark of trees. This was unlike the media used by other civilizations, such as clay in Babylonia and papyrus in Egypt (Struik, 1987).

The Harrapan was an impressive civilization established in the Indus Valley between 2,300 and 1750 B.C.E. It was stimulated by constant contact with Mesopotamia. It was later absorbed by the Aryans who invaded the Ganges River valley between 1500–1000 B.C.E. Their language, Sanskrit, was the vehicle for much literature and science produced in the period including the *Rig Veda,* which gave mathematical information that was conveyed during the telling of religious myths (Cooke, 1993). Much of the developments during the Aryan occupation were secular.

Indian culture flourished. By the sixth century two impressive religions came into being: Buddhism, named for Guatama Buddha who lived 563–485 B.C.E.), and Mahavira, founder of Jainism made contributions to mathematics. Religion heavily influenced Indian civilization. Many mathematical problems were described in linguistic terms. Indian mathematicians used abbreviations or the syncopated style of algebra.

Mathematics flourished in India during the sixth to fifth centuries B.C.E. Two documents came into being during this time: the *Surtras* and *Siddhantas,* which built on knowledge of earlier generations (Cooke, 1993).

As in other civilizations, mathematics in India was based on practical considerations, that is, what happened in life. The Indians, however, were a religious people, and their mathematics also reflected this preoccupation. Mathematics during the Sutras period was related to the rules of religious life, and many were written in verse. Both number theory and geometry were based on religious rituals. For example, a Hindu home had three altars of different shapes and prescribed area. A related mathematics problem showed integers written in triplets (3, 4, 5) (5, 12, 13) (8, 15, 17). This problem led to a kind of problem later deduced by Pythagoras.

Also based on this problem of the altar, transformation geometry was invented (e.g., a square within a circle). The Indians also knew irrational numbers. Because Indian civilization was religious, mathematics was based on metaphysical concerns with matters dealing with the Infinite. For example, the Jania classified numbers as enumerable, unenumerable, and infinite, and one dimensional, two dimensional, and infinity. There was also the classification of beings. For example, animals were classified in accordance with the five senses. In the Bragahata, written around 300 B.C.E., the author asked, "How many philosophical systems can be found by a certain number of doctrines?" This led to the principle of combinations. The Indians also used the principle of computing coefficients. In 1881 a manuscript was unearthed in Bakshali, an Indian village. It contained problems in linear equations based on real-life problems. In the fourth century C.E., the *Suddhantas,* a treatise on astronomy, was compiled. It demonstrated a well developed system of trigonometry in which the word *sines* was coined.

With the conquest of Alexander the Great in 300 B.C.E., India became part a vast polyglot empire, and there was much interchange with other civilizations including the Persians and Greek. During the first century, Islam made its appearance in India. It copied much of Indian mathematics including the decimal system and the zero.

The next period of the development of Indian mathematics, between the fifth century and the twelfth century C.E., was dominated by the works of Aryabhata, Brahamagupta, and Bhaskara. Aryabhata's book called *Ary-abhatya*, was discovered by an Indian scholar, Bhau, in 1864. It was written in verse with 123 stanzas and was concerned with trigonometry, measurement, and geometrical problems that dealt with triangles, circles, spheres, and an approximation of pi. The concept of sine as half cord was also used, and the work contained a table of sines. In algebra there were problems of the arithmetic progression and finding the sum of terms.

Brahmagupta (628 C.E.) is perhaps the most famous Indian mathematician. He wrote a thesis titled *Brahmasphutuse.* His work was translated to Arabic and it was through his work that astronomy was brought by the Muslims to Europe. He had rules for computing fractions, square roots, diameter of a circle, and quadrilaterals. In algebra, Brahmagupta used the syncopated style invented in Babylonia. For example he used *YA*" to mean X, *KA* for Y, and *BHA* for product (Eves, 1969).

The third outstanding Indian mathematician is Bhaskara, who lived in the twelfth century C.E. His work was *Siddhanta,* which discussed problems in algebra. A problem called *Lilavate* dealt with combinations and permutations. He also invented the rules for solving quadratic equations using radicals. In astronomy, he gave the rules for solving trigonometric problems using sines. Indian mathematics utilized linguistic development to an art form. They devised logical arguments for visual demonstrations to expand concepts. Both Brahmagupta and

Baskara used visual forms. For example, they were able to show that when XY is not 0, an infinite number of solutions can be found and they used a visual method to compute squares.

Because linguistic empiricism and visual representation were used instead of logic, the results were validated by checking, which was a kind of proof. For example, summation of an infinite series is an Indian invention based on visual demonstration without reference to axioms. Perhaps this is the weakness of Indian mathematics, which makes it difficult to duplicate. A typical problem was "a cricket field has an area of 144 sq feet. The length exceeds the width by 10 feet. Find the length and width" (Wrestler, 1989). Although the Indians knew the form $X^2 + 2X + 19 = 144$ and could demonstrate it empirically, they could not formulate the reasoning symbolically. Indian mathematics also contributed the square root invented by Sulbasutra (800–500 B.C.E.) and later explored by the Greeks as irrational numbers.

The Greeks: Mathematics Based on Deductive Logic

During the last century B.C.E., the economic and political climate created new demands for mathematics. The balance of power had shifted from Babylonia and Egypt to Greece, which emerged as the dominant imperial power with colonies that lasted from 500 B.C.E. to 320 C.E. In 300 B.C.E., when Alexander the Great of Greece conquered vast territory extending from Asia Minor, Egypt, Mesopotamia, Persia, Afghanistan to the borders of India, the Greeks established hegemony over an extensive polyglot empire that they promptly proceeded to integrate and acculturate.

Greece became the intellectual capital of the known world and their mathematicians began by copying mathematical forms such as rhetorical algebra from Babylonia and Egypt. At first, Alexandrian Greeks had empirical preoccupations such as solving problems relating to commerce and the division of labor. An Alexandrian Greek, Diophantus (75 or 250 C.E.) used this rhetorical form of algebra. A typical method of presenting the volume of a cube was "the solid whose sides are first, second, and third unknown quality" (Eves, 1969). Some scholars attribute the invention of algebra to Diophantus, who invented the syncopated form of algebra that uses abbreviations and the parametric methodologies that present two sides of an equation. Euclid's work parallels the Babylonian algebra for example the form,

$$(a + b) = a^2 + 2ab + b^2$$

Some Greeks had interest in practical methods but used deductive methods. According to historian Herodotus, Thales had both scientific and practical interests. He predicted an eclipse of the sun in 585 B.C.E. and helped Cresus to divert a river so that his soldiers could go across. On the theoretical side he demonstrated vertical angles of two intersecting lines and characteristics of isosceles triangles, the circle, and congruent triangles. It was Thales of Melitus who invented demonstrative geometry. Another Greek who had interest in practical activities was Apollonius (255 B.C.E.), an astronomer, invented theorems based on conic sections, parabola, and ellipse for use in his professional activities.

The later European Greeks abandoned the empirical methods of their predecessors and proceeded to show interest in geometry and logical deduction. Copying from the Babylonians Euclid expressed algebra in geometrical terms, for example, $A^2 + 2ab + B^2$ was expressed geometrically. The economic and political climate in Greece was more international and sophisticated. Their involvement with foreigners and preoccupation with commerce made them become rich. Released from the struggle for economic survival, and having attained great riches, the Greeks could afford to use leisure in the development of culture and in intellectual pursuits. Non-Euclidian mathematicians had a deep and abiding intellectual curiosity. They were interested in the *whys* rather than in the *hows* of mathematics, without regard to practical application. They became more sophisticated than the Babylonians and insisted on theorems, formal deduction, and proof. This was not unusual as the there was a philosophical intellectual climate in Greece and the Aristotlean rationalistic methods were transferred from philosophy to mathematics. The Greeks believed that the purpose of mathematics was to explain man's place in the universe and that solutions should start with axioms, a principle that still exists today. They invented deductive or demonstrative geometry.

The Pythagorians, a group of mathematicians and philosophers, made significant contribution to mathematics, developing number theory which they classified into primes, composites and perfect. Pythagoras (540 B.C.E.) himself, formed a school, and he and his followers traveled to Egypt and Mesopotamia. Pythagoras himself created concepts in geometry points, lines, planes, solids velocity, and of course his famous theorem concerning any right-angled triangle: the square of the hypotenuse is equal to the sum of the squares of the other two sides.

Other mathematicians were just as productive. The famous Euclid published 465 propositions on solid geometry and number theory, and Archimedes (287–212 B.C.E.) derived methods for calculating measurement of a circle, the parabola, and sphere and cylinder.

The Romans: Administration and Technology

With the death of Alexander the Great, a new power emerged in the Mediterranean basin. This began in 2,000 B.C.E. with the Latin town of Rome, which became a Republic in 510 B.C.E. and an Empire in 146 B.C.E. The Roman Empire was a polyglot body that extended from Byzantine in the East to Western Europe as far afield as Britain and South to North Africa. The Roman Empire enjoyed the wealth accrued by imperialism and inherited the cultural legacy of the Greeks; hence culturally it was initially the Roman-Greco Empire.

The Roman Empire is distinguished by its great military accomplishments, its administrative bureaucracy, jurisprudence, and outstanding advances in technological advances including architecture, geography, engineering, and construction of aqueducts, roads, towns, bridges and tunnels, and military fortifications, all of which were inspired by mathematics. Engineering was the major strength of the Roman Empire.

In the second century the Roman armies ceased to roam and were settled in camps. This gave rise to fortifications that were constructed to give security, such as the Hadrian Wall in Britain and forts along the Rhine and Danube Rivers. Similarly the Romans developed a network of roads throughout the Empire. Major towns were built at strategic points in a manner similar to large scale engineering projects. Cooke (1993) explained that each town was laid out at the intersection of two perpendicular roads.

There were lots called *centenurioe* that were laid out like a Cartesian plane and units called *dextra decumani*. Surveying was done in connection with road building and an instrument called the *groma* was widely used. A damaged sample of a *groma* that was used in this project has been recovered and is now in the Museum of Naples.

Geography was a major preoccupation of the Romans. Knowledge of geography was needed to administer the Empire. Cooke (1993) explained that Caesar Agustus commissioned Agrippa to create a map of the world. This was done and placed on the wall at Via del Corso, now destroyed. The mapmaker had to have knowledge of plane surface and the curvature of the earth. Ptolemy also made a map that is now extant.

Architecture was perhaps their greatest contribution to civilization. An examination of Roman architecture reveals extensive use of mathematical figures. The arches, bridges, tunnels, buttresses, and amphitheaters all reveal trapezoids, semicircles, triangles, and the use of pi. Roman mathematicians created mathematics for utilitarian purposes, unlike the Greeks who were cultural creators.

Diophantus (75–250 C.E.) was a Roman-Greek who lived in Alexandria, Egypt. His masterpiece, the *Arithmetica,* was a treatise that described polygonal numbers and indeterminate equations, that is, those with two or more equations, several variables, and an infinite number of rational solutions. His method of solution copied the Babylonian parametric style. Diophantus was the one who invented syncopated style, which was a step beyond rhetorical style and nearer to modern algebra

Claudius Ptolemy (85–165 B.C.E.), a Greco-Roman who also lived in Egypt around the second century, was an astronomer and mathematician who developed the heliocentric system of the universe. This system dominated learning for one thousand years until Copernicus and Galileo proposed ideas that were directly opposite. In his book, the *Almegest* (150 C.E.), which still survives today, he discussed trigonometry in detail and the circle as having 360 degrees. His table of sines gave the length of a cord, "subtended from a circle as 60 units, having an arc of 0 degrees to 120 degrees through increments of 1/2 to 120 by steps of 1/2 degree was equal to sine ¼ degree to 90 degrees by 1/4 steps."

Ptolemy also created a map of the world. Pappus lived during the reign of Diocletian (285–315 C.E.). He wrote the *Collection*, which consists of commentaries by Euclid, Ptolemy, Apollonius, and Archimedes, on subjects such as arithmetic, geometry, and construction. Hypatta (355–414 C.E.) of Egypt was a female mathematician who contributed geometry and astronomy. M. Iunius Nipsus was a surveyor who wrote an article in a two-volume work titled *Corpus Agrimensorum Romanorum* that was translated into German in the nineteenth century. The article described a method of surveying that gave directions for measuring the width of a river using congruent triangles.

Anacius Boethus (475–524 C.E.) lived in the twilight of Roman civilization and was an educator who coined the terms *Trivium* (dialectic, rhetoric and logic) and *Quadrivium* (astronomy, arithmetic, geometry, and music) to describe the Seven Liberal Arts. He wrote the text, the *Institutio Mathematica,* which became an authority on mathematics that was used for one thousand years. His work also dealt with the Pythagorean Theorem.

Gerbert (999 C.E.) was a French monk who later became Pope Sylvester. He developed a version of the abacus, a counter with twenty-seven columns called *apices* and no zero counters.

The Successors of the Roman Empire

The civilizations of China and India lasted continuously at a high peak until around the thirteenth and sixteenth centuries C.E. respectively. Greece had been absorbed by the Roman Empire, which conquered it during the Punic Wars. The Roman Empire lasted from the founding of the city of Rome by the Latins in 1,000 B.C.E., until 450 C.E., when it fell to the Germanic tribes. There followed three successor empires: The Byzantine, Islamic, and Christian Western Europe. A vibrant civilization continued in Byzantium, which preserved the traditions of the Greek and Roman Civilizations, and reintroduced classical learning into Europe during the Italian Renaissance of the fifteenth century. A dynamic and influential civilization flourished in Islamic Empires but learning reached a nadir point in Western Europe, a period called the Dark Ages.

Islam: Synthesized Greek Deductive Logic and Indian Linguistic Numbers

Islam had its genesis in the Bedouin tribes of the Arabian Peninsula with its founder Muhammed. It was a theocratic civilization that synthesized civil government and religious practices. Between 632 and 750 C.E., the Arab armies conducted warfare extending from the Arab peninsula, North through the Persian Empire to the borders of the Byzantine Empire, East to India and China, and West across North Africa and from Africa north into Spain. In effect the Islamic Empires controlled the entire Mediterranean basin that had once been dominated by the Roman Empire. Classical Islam lasted until 1200 C.E.

The Islamic Empires absorbed the technology, ideas, and intellectual traditions from the civilizations it conquered while developing a civilization of its own. The Empire was unified by the use of the Arabic language and the Muslim religion, which were universally adopted. The adoption of cheap writing paper from China greatly facilitated communication throughout the Empire. Paper was obtained inadvertently from China, through a Chinese prisoner in Samarkland in 751. It was adopted in Baghdad in 793, in Egypt in 900, and Spain in 950. Islamic rulers called *caliphs* fostered scientific research, and the palace promoted scholars.

Islamic scholars synthesized the methods used by people they conquered. When they routed Egypt, the Islamics discovered in the Alexandrian libraries the works of Greek scholars such as Archimedes (287–212 B.C.E.), Apollonius (225 B.C.E.), and Euclid (300 B.C.E.) as well as Greco-Roman scholars such as Diophantus and Ptolemy. Because Islam tolerated minorities, it accepted the Syrians, who were heirs to Greek civilization. From them, Islam obtained the works of Aristotle, which influenced their philosophy and theology. Islam translated these works into Arabic, adding commentaries. Scholars such as the philosopher Averroes, and the scientist and physician Avicenna made valuable contributions to learning. Islam also absorbed medicine and astronomic traditions from Persia and developed trigonometry, which they used to solve problems in engineering, navigation, and physics.

On conquering India, they availed themselves of mathematics from which they copied the number system called Arabic numerals that is still universally used today. They copied the decimal system and the use of the zero from India. Islam translated Greek and Hindu works into Arabic.

They synthesized the method of using geometry to solve algebraic problems used by the Greeks and the number system of the Indians. At first Arabs used these methods by writing out problems in words but then adapted Hindu symbols. Later Arabs reverted to rhetorical style in which problems were written, showing the influence of Greek methods. By synthesizing the deductive logic of the Greeks with the linguistic empiricism of the Indians, the Arabs created a viable and dynamic mathematical model that formed the basis for further development in Europe especially through Spain and Italy.

Arabic scholars also created works of their own. The mathematics of Omar Khayyam (1050–1123) and Muhammed ibn Musa Al-Khowarizmi (Mohamnmed, son of Moses) made significant contributions to mathematics.

Al-Khowarizmi (al-jabr, 825 C.E.) gave his name to the word algebra. The name is derived from his book, *Kiab fi lajabar wal muqabula,* which, when translated into Latin was called *Aljabr* and gave rise to the word *algebra.* Al-jabr describes the process for solving equations meaning (1) transportation or gathering and summing all like terms on each side of the equation and (2) cancellation, which means canceling like terms on opposite sides of the equations. Because the idea of collecting like terms did not occur in either Greek or Indian works, it appears to be his invention. Al-Khowarizmi copied the rhetorical style of Diophantus, to which he added geometrical style of the Greeks to solve equations. He did problems in area and volume and used both numerical and geometric methods to solve quadratic and linear equations. Al-Khowarizmi used mathematics to solve practical problems such as legal problems and problems of inheritance (Cooke, 1993).

He also wrote books on astronomy and mathematics, and it was his work *Algorithmide Numerto Indorum* was the main instrument by which mathematics entered Western Europe, and which was used by later medieval mathematicians. Among the concepts he explicated, was the Hindu decimal system. Because Algebra came to Europe from the Arabs, the method that dominated mathematics until recently had little deductive foundation, emphasizing method rather than logic. Also geometry emphasized postulates and theorems.

Thabit ibn Qurra (836–901) did his research into the mathematics of Apollonius, Archimedes, Euclid, and Ptolemy. Based on these works he created original mathematics in number theory and geometry. Using similar triangles, he created a variant of the Pythagoras theorem.

Abu Kamil (850–900) wrote a commentary of Al-Khowarizmi's works in which he gave the basic rules of algebra. He used the terms *square* and *roots* and *radical* based on both Greek and Indian ideas of the side of a square. Kamil created a classification of quadratic equations and methods for solving them. He used modern notations of equality. His work contained sixty-nine problems based on practical mathematics.

Omar Khayyam (1050–1123) is known to the west for his poetic work, the *Rubaiyat.* He was also an influential mathematician of Persian ethnicity. His work *Algebra* shows algebra that is a synthesis of number theory and geometry and describes a classification of quadratic and cubic equations that are solved by geometric methods. Based on Diophantus's methods, he did not see the possibility of a fourth power, and following Euclid, he represented the unknown by a line segment. Khayyam relied on Al Khowarizmi's methods of solving equations.

Western Europe and the Revival of the Intellectual Climate

Early Middle Ages. The fall of Rome created in Europe a political partition of Eastern and Western Europe. Scientific thought had always come from the East, and with the fall of Rome the agricultural feudal west was isolated and then maintained a static civilization called the Dark Ages. During the early Middle or Dark Ages, the Church maintained as best it could any semblance of civilization left by the Roman Empire, as clergy and laymen continued to keep alive the Roman-Grecian civilization. Mathematics was mostly mensuration and arithmetic. Among the mathematicians were a layman named Boethus and a monk called Gerbert.

High Middle Ages began with the revival of trade and commerce, the revival of trade with the East and the reconquest of Spain from the Muslims. By the beginning of the twelfth century, the Arabs had made themselves masters of the Mediterranean basin and they dominated the trade routes that had facilitated the trade of the Roman Empire. They dominated the region from Italian Sardinia, Corsica, Sicily, to Palestine, Syria, North Africa, and Southern Spain. The recovery of Europe was on two fronts, in the West in Spain and in the Italian cities.

The Arab conquest of North Africa and Spain spurred Christians from all over Europe to respond in a momentous movement called the Crusades. The Crusades did not achieve its intended outcomes of driving the *infidel* from Jerusalem. There were, however, unintended consequences that were so profound that they changed the course of history. This new phenomenon included the revival of economic activities, the growth of commerce, towns and urban civilization, a bourgeois society, and the revival of intellectual activities, most important of which were the universities. The earliest cities to begin trading activities were the Italian cities of Genoa and Pisa.

Traders established business relations with the Eastern Mediterranean cities where the Byzantine civilization that had inherited

Greek civilization was still flourishing. Later French and Central European cities followed suit.

Following in the footsteps of traders were scholars who were able to study Greek civilization at first hand. Another channel through which Europeans forged links with classical antiquity was through Spain. Christians finally captured Toledo in 1085, drove the Islams from the city and in 1212, and finally drove the Muslims from Spain. It was in Spain that Christians interacted with Islamic civilization. Scholars translated Greek writings from Arabic into Latin, and Greek knowledge was assimilated through Arabic.

Leonardo of Pisa or Fibonacci (1180–1250) was one of the merchants who traveled to the Orient. His real name was Leonardo of Pisa, called Fibonacci, and was a member of the house of Bonacci. He collected pertinent information about arithmetic and algebra, which he recorded in a book, *Liber Abaci,* or *Book of Abacus* (1202) on his return to Pisa. This book was one of the channels by which the Hindu-Arabic numeral was introduced into Europe. Leonardo also wrote *Practica Gometra* (1220), which gave a description of the geometry and algebra, which he had learned. He also made several original observations but also made references to Al Khwarizmi's equation,

$$X^2 + 10X = 39.$$

This led to the famous Fibonacci series (0, 1, 2, 3, 5, 8, 13, 21) in which each term is the sum of the preceding terms.

Jordanus Nemorarius based his work on the Algebra of Al-Khowarizmi. In Nemorarius's books, *Biblionomia* and *De Numeris Datis,* he demonstrated that if the sum and difference of two numbers are known, then the numbers can be found.

The Spanish channel also produced the Indian-Arabic numeral, the oldest extant book being the *Codex Vigilantus* (976). Italian merchants were the first to use Arabic numerals in their accounting records. It slowly replaced the abacus, which had been widely used by the Russians, Chinese, and Japanese. Trade and its practical requirements were the engine that drove mathematics. It was used in bookkeeping, computation, navigation, surveying, and astronomy.

After Leonardo, Italian merchants continued to show interest in practical mathematics. Most of these mathematicians were technicians such as instrument makers and printers. Johanesnes Muller of Konisberg, an instrument maker and printer translated classical manuscripts, and a painter, Vasari, demonstrated how to find poof in geometry. Vasari (155 C.E.) wrote *Lives of the Painters* in which he discussed solid geometry. Another achievement was the study of solid bodies in perspective by Brunelleschi. The invention of printing in 1454 by the painter Johannn Guttenberg led practical mathematics to its final conclusion.

Speculative mathematics was not for practical men. This was left to churchmen such as St. Augustine, and St. Thomas Aquinas. University scholars sometimes joined practical men in the pursuit of mathematics. Some scholars were interested in satisfying their own curiosity, sometimes for practical reasons and sometimes to make a contribution to the canon. It was during this time that the writings of Aristotle were reconciled with Christianity by St. Thomas Aquinas. Generally the achievements of the past remained the inspiration of men of the fourteenth century; there was little innovation, and the status quo was maintained.

The Renaissance: Mathematics Applied to Art and Architecture

In 1453, the Turks captured Constantinople of the Byzantine Empire. This occasion ushered in the period of "rebirth" of learning in Western Europe known as the Renaissance. The Renaissance was characterized by Humanism, the study of man rather than of God. This was, however, the study of man in classical antiquity, past Greek and Roman models, for inspiration. The beginnings of capitalism also occurred in the Renaissance. Modern science reared its head in the form of the mariners' compass. Johannn Guttenberg invented printing. Art flourished. The Renaissance gave birth to new visions, ideas, and achievements.

Thus after a period of quiescence, mathematics returned to Europe in full force. Mathematics during the Renaissance was built on the tradition of the practical use of mathematics, particularly the use of machines that had become prevalent during the past century. The merchants, in need of money for trade, and princes with the desire to perfect the tools of warfare had inspired the construction of new machines that were dependent on mathematics for their perfection. These beginnings and the mathematical base that influenced their use were perfected during the civilization of the Renaissance. Because the

Renaissance was a cultural movement, mathematics of the period was related to art and architecture and dealt mainly with perspectives, the laws controlling the projection of objects on screen. Leonardo de Vinci and Raphael were students of perspective. Demand for practical mathematics continued unabated, however.

In addition to astronomy and commerce, the leaders of emerging nation-states such as Italy needed mathematics to solve problems in navigation, engineering, and surveying as well was in the military. Engineers who were called upon to create large public works needed mathematics and for military construction.

Regiomontanus (1436–1476), whose real name was Johannes Muller Konisberg, was an instrument-maker and printer. His work influenced algebra and trigonometry. He created a table of trigonomic ratios and tables of sines. Trigonometry also made strides. It inspired the works of Copernicus, Tyco, Brahe, and Kepler, and brought into focus new ideas about the nature of the universe.

Scipio del Ferro (1456–1568) invented a theory, missed by the Arabs, that led to the solution of the cubic function and by extension to all general solution of equations. His method was rediscovered by Tartyaglia in 1535, who enjoined Cardona, a mathematician, to keep it a secret. The latter did not.

Early Modern Europe and the Development of Symbolic Mathematics

In Europe in the sixteenth century, mathematicians made great strides in proffering great ideas that were to begin a revolution in mathematics. The most important development in algebra was the standardization of methods of presenting algebra. Around 1500 the rhetorical or syncopated forms gave way to symbolic language. As for example Robert Recorde (1519–1558) used the equal sign to replace "equale" and Christoff Rudolff (1500–1545) used the square root sign to replace the word *aradix* or *root*. Girolamo Cardona (1501–1576) replaced "cubus p 6 rebus aequalis" with X63 + 6X = 20 (Baumgart: 1969).

Cardona was a transitional figure who made an innovation that made mathematics more than just manipulation of equations and relegated the application of ingenuity in solving algebraic equations to the past. He published the solution to del Ferro's theory and also made a brilliant discovery of his own: to reduce the quadratic equation to a cubic function, for example,

$$X^4 + 6X^3 + 36 = 60X \text{ is reduced to } y^3 + 15y^2 + 36y = 450.$$

He was unable to reduce complex numbers, a difficulty that was solved by Raffael Bombelli in 1572.

Bombelli (1526–1572) was concerned with the cubic function that could not be reduced. To solve this problem he invented the name "plus of minus" to mean the negative of a square root of (-1) and minus of minus one. He suggested that these two numbers were inverses of each other; they were first added and then subtracted.

Francois Viete (1540–1603), a lawyer in the French court of Henry IV, revolutionized the standardization of the presentation of mathematics. He solved the Cardona's cubic function by using trigonometry. He stated that the "left-hand number was equal to sine in terms of sine 450." Viete was the pioneer of representing algebraic terms by using letters. Viete explicated in his *Logistica Speciosa* how mathematics could be systematic. He demonstrated how theoretical properties could be applied to an equation. He introduced the idea of a coefficient, roots, constant, and degrees of polynomial (Baumgart, 1969). For example he translated

$$1QC - 15QQ + 85C - 225Q + 274N \text{ aequatur } 120$$

as

$$X^3 - 15X^4 + 85X^3 - 225X^2 + 274X = 120$$

Viete introduced rules for the powers of the binomial and the relationship between the root and the coefficient and used vowels to describe the unknown and describe data. He further used the plus and minus signs in their present forms (Struik, 1987). Viete brought the calculation of pi a step further than Archimedes by expressing pi as an infinite product. This improvement in notation made algebraic calculations more accurate. This stage of notation was made definitely possible by new methods of writing, that is, the use if the Hindu-Arabic numerals. In effect Europeans had mastered and surpassed the mathematical achievements of the Arabic world. Viete's work marked the beginning of modern algebra.

The European Enlightenment: Influence of Science and Technology

The seventeenth century saw the rapid industrialization of Europe, the development of machines, and the application of science to manufacturing. Galileo's spirit of inquiry gave birth to many inventions in general. The European Enlightenment, the intellectual movement of the seventeenth and eighteenth centuries, imprinted its main characteristics of rationalism, science, humanism, empiricism, and rational interpretation on the mathematics of the period. Although Bacon (1561–1624) and John Locke (1632–1704) had proposed empiricism as an epistemological method, it was rationalism, the traditional philosophy that Rene Descartes (1596–1650) applied to mathematics that became famous.

Descartes, a philosopher and mathematician, synthesized algebra and geometry using the Cartesian plane that was named for him. He created a new way to write equations:

$$X^2 - 6X + 13X - 10 = 0$$

In essence the main contribution of the Europeans of this age was to transform mathematics from a syncopated style to a discipline based on symbols and exact calculations. Perhaps the main inheritance of mathematics of the period is the difficulty that children experience in making the transition from empirical mathematics to applying theories.

Descartes took his cue from the Greeks, especially Platonic Idealism. He conceived of mathematics as a body of infallible truths to be discovered by reason. The quest for absolute truth and "clear and distinct ideas" was the basis of his mathematical investigation. Based on his epistemology of rationalism Descartes, philosopher and mathematician, deduced that universal truths could be discovered from a few axioms. Thus he invented analytic geometry. Whereas the Babylonians and Egyptians had used this type of mathematics in surveying and construction, and the Greeks had used it in mapping, it was Descartes who applied algebraic symbols to the coordinate plane named for him (Eves, 1989).

His invention, in effect, synthesized geometry, algebra, and trigonometry, making it possible to solve challenging problems, hitherto unexplored. By using analytic geometry, practitioners may use algebra to solve geometric problems based on real-life. He also created a new way of writing algebraic equations thus,

$$X^2 - 6X + 13X - 10 = 0$$

While Descartes and Fermat (1830) explored the area of geometry, their compatriot Pascal (1662) made his contribution in the new field of projective geometry.

Undoubtedly the most significant mathematicians of the late seventeenth century were Isaac Newton (1646–1727) and Wilhelm Leibniz (1646–1716), who independently invented calculus. The genius of Newton synthesized the method of science and the rationalism and applied it to mathematics. Hypothesizing that the force that made an apple fall to the ground was the same force that made the moon bend into orbit, he developed calculus to prove it. By these methods infinitesimal calculations pertaining to the problems of dealing with curves and surfaces could be easily solved. Newton used it in physics and mechanics and it has been found to be a useful tool in engineering, navigation, and surveying. It made space travel possible. Thus Newton demonstrated that mathematics could be used empirically as well as rationally. Also using science, Leibniz developed calculus to measure accelerating and decelerating motion.

Mathematics: Number Theory and the Military

Between 1600 and 1800 the main algebraic inventions were in the area of number theory. The main concepts that entered the field were logarithm by John Napier used for calculating (1550–1617), probability by Pierre de Fermat (1601–1665) and Blaise Pascal (1623–1662), the binomial theorem by James Gregory (1627–1675) and Isaac Newton, and actuarial science by Abraham de Moivre (1667–1754). Several mathematicians, such as Larange (1736–1813), were busy trying to find a general solution for polynomial equation (Baumgart, 1969).

Mathematics also had a utilitarian reality during the eighteenth century. It was applied to practical situations such as three dimensioned objects, fortifications, and the military. Victor Poncelot (1788–1867) created projective geometry and used it to escape from prison. According to Struik (1987), "The excellence of the French navy was due to the fact that in construction of frigates and ships of the time, the faster shipbuilders were partly led by mathematics theory."

During this period, too, many military schools were founded, the first being the *Ecole Polytechnique* of Paris in 1794, which became a prototype and inspiration to others, including West Point in the United States.

Mathematics in the Twentieth Century: Sets, Axioms, and Deductive Logic

The nineteenth century saw the acceleration of growth of capitalism, industrialization, and specialization. The international nature of mathematics gave way to nationalist developments. Mathematicians were no longer a repository of the leisured class. In this era they became respected researchers, theorists, teachers, and technocrats. Theoretical mathematicians were preoccupied with the search for a general solution for the polynomial equation.

Those who built on the work of Larange were Galois (1811–1832), who invented factor theory; Henrik Abel (1802–1829), who gave the world permutations; and Cauchy (1789–1857) and Leopold Kronecker (1823–1891) derived sets and inverses. Following Carl Gauss (1777–1855) manipulation of complex numbers, William Hamilton (1805–1865) invented his theory of vectors. The mathematical stage was characterized by the influence of *quarterions* and the commutative and distributive laws. In exploring this concept, mathematicians began to appreciate that geometry was applicable to algebra, and they began to experiment with new systems called axioms.

The early twentieth century saw mathematics being applied to industries, insurance, and engineering. It became mostly a theoretical discipline associated with research and universities. Most modern mathematicians were Europeans, but as the years progressed, mathematics became more international and pragmatic.

In the late twentieth century, mathematics again underwent a change in direction. The earliest practitioners were inspired by natural phenomena, and mathematics was used as an instrument to describe reality. Today mathematics is a discipline and a science in is own right, deriving ideas from within itself. Mathematics in the twentieth century became more abstract and based on definitions rather than on intuition and common sense. Every theorem has to be rigorously interpreted and proved as mathematics has to present a point of view. Whereas the

Babylonians had used computation for practical purposes, modern mathematics has developed number theory based on research and proofs, the era of axioms had arrived.

The most significant invention of this era was set theory and symbolic logic and exploring the interrelationships among types of geometrical structures and algebra. Theories of sets and integrals first defined by Archimedes (287–212 B.C.E.) have been in this century rigorously defined by such proponents as Bernhard Riemann (1826–1866), who investigated functions and derivatives.

Bertrand Russell (1872–1918) and Alfred North Whitehead (1861–1947) presented to world with symbolic logic in their *Principia Mathematica* (1910). New concepts such as integral, topology, and matrix have been introduced in the twentieth century. Mathematics has become inextricably linked to science and the method of investigation has become axiomatic deductive reasoning.

Mathematics and Technology

The aftermath of World War II profoundly affected mathematics. In 1946, the first computer arrived on the scene. Given life in the logic of the Greeks, its growth continued through the works of Wilhelm Schichard (1632) and Pascal and Leibniz. Following these beginnings, Joseph-Martie Jacquard (1808) invented a punch card system and Charles Babbage (1833) an analytic engine. American Hollerith (1890) standardized the punch card. German Konrad Zuse (1934) continued the binary system, and Vanevar Bush and Norbet Weiner (1939) of the Massachusetts Institute of Technology created a computer for evaluating integrals. An Englishman, Alan Turing (1936), devised a logical machine, and a practical computer was created in 1937. In 1937 Howard Aiken in collaboration with International Business Machines (IBM) created the Mark 1. The first electronic computer UNIAC was completed in 1946.

Mathematics had reached its full flowering in the modern computer, which is based on the Boolean algebra of George Boole (1815–1864) and matrix algebra of August Heisenberg (1925). These algebraic methods were useful for describing phenomena in quantum terms. Consequently new fields of mathematics opened: linear programming, operations research, set theory, and probability. Mathematics had given the world the computer, which could be used for solving everyday problems in academics, industry, science, business, and indeed in every walk of life. Algebra, such as the matrix, is based on real-life situations, and endless potential is waiting to be explored. In 1975, Bill Gates founded a corporation called Microsoft and created software that could perform mathematical functions. Technology that was born in mathematics is threatening to make it anachronistic. Effective users of technology, however, must first know mathematics.

The Teaching of Mathematics in Real-life Situations

The historical development of mathematics has demonstrated that the subject is related to economic and social contexts. At every period of its development, new concepts were added, and the discipline was related to prevailing ideologies and practical needs. More than ever, today society is characterized by technological developments as its essential lifestyle. At every stage mathematics is required to solve real problems. Indeed common existence is impossible without the application of mathematics. Whether it is obtaining a mortgage loan, purchasing annuities, constructing bridges, or houses, orbiting in space, communicating across national boundaries on the Internet, or playing the lottery, a knowledge of mathematics enables one to operate effectively.

The teaching of mathematics in a practical context should be basic to any pedagogy. Real-life situations arouse the curiosity of students, challenge them to solve problems, and invite them to explore and create new mathematics concepts that may be used for pragmatic ends. The idea of continuous progress depends on the existence of dynamic creators of new mathematical ideas. Mathematics is to be seen not only as part of the cultural heritage with cogent formulas and principles to be appreciated, but it is also an instrumental body of knowledge, which is basic for survival in this technological age.

Teachers may give a wide range of assignments such as research on the historical development of a concept such as the zero or rational numbers. They may require students to reinvent a concept and apply it empirically in real-life situations. For example, a quadratic equation represents a rectangular surface; students may be required to apply this knowledge to the construction of a display window and to factor equations. Students may also be asked to pose problems and represent an event mathematically such as the sine and cosine functions to calculate the distance from New York to Frankfurt.

Conclusion

Historical analysis has been the basis for the theory that mathematics should be related to life situations. Mathematics was invented in the great civilizations in Babylonia, Egypt, Aztecs, Chinese, Indians, Greek, Romans, and Western Europe as a response to social and economic needs and problems of the age. In the earlier civilizations, solutions were mostly empirical, which gave rise to deductive and theoretical methods in the modern age.

Methodology had been related to the ideologies of the age, from concrete representations in China and linguistic and visual empiricism in India to the symbolic and axiomatic of the twentieth century. To a large extent the present age is a composite of the cross-fertilization of experiences and assimilation of the best mathematical ideas, born in past cultures and civilizations. Educators would do well to look to the historical development of mathematics and its relationship to life situations to derive suitable methodologies for implementation in the educational system. As mathematics educators base their epistemology on past and present models, it behooves all mathematicians to foresee that the best is yet to be.

ASSIGNMENT

Read the preceding chapter and the references that follow and then answer the following questions.

1. Write a description of the history of algebra or calculus.

2. Describe instances of how mathematics was used to solve practical problems.

3. For what reasons was trigonometry invented? Give examples.

4. Write a description and rationale for the main mathematical signs and terms for example zero, rational numbers, pi.

5. Write a description of how geometry was used in architecture and art.

6. Give an account of the importance of mathematics in the military.

7. Write an account of the development of technology.

Figure 1-1. Mathematics in the Design of Chartres Cathedral, France

Figure 1-2. The Mayan Calendar

Chapter 2

History of Mathematics Reform in the United States since 1950

During the last five decades of the twentieth century, American schools have been characterized by what has been called the reform movement in mathematics, each period reflecting the economic and sociopolitical ideologies of the decade. Each period was a reaction to the previous one. The subject-centered focus of the Cold War curriculum of the fifties was opposite to the ethnic and gender-related, social equity goals of the turbulent sixties; was more consistent with the conservative back-to-basics, competency-based learning of the seventies and the academic excellence of the eighties; and the national goals, content-standards, and restructuring of the nineties. The twenty-first century has begun as an outgrowth of the nineties, especially with standard-based content and the global economy. Generally, reform led to changes in curriculum content, which in pedagogical terms is translated to different teaching methodologies in the classroom.

The New Mathematics

Mathematics in the 1950s was the first "back to basics" era. It was inspired by "Cold War" politics. When on 1 October 1957 the Soviet Union launched the first earth satellite, a small 184-pound sphere that orbited the earth once every ninety minutes, the United States was concerned that its national security was threatened, as it had lagged behind in technological development (Huetinck & Munshin, 2000). Less than four years later, the Soviet Union launched the first spaceship with a human pilot.

America was not able to match any of these feats until several months later, and the general belief was that the Soviet Union was ahead in space (Cooke, 1997). According to Carlson (1966), "No single event had greater impact on public opinion or more strongly reinforced the conclusion that the American school system was failing in the race with the Soviet Union for world leadership." It was the beginning of the space race that was to alter profoundly the relationship of the federal government to the nation's schools and universities (Brademas, 1987). Educators of the time attributed this failure to the pedagogy of the previous progressive era. They charged that the school system had abandoned the formalistic, ritualistic, subject-centered curriculum and replaced it with "soft pedagogy." Policy-makers and educators generally agreed that nothing but complete reversal to hard pedagogy would supply the United States with essential scientific potential that would make the United States competitive and even superior to the Soviet Union (Bereday, 1964). The conservative, elitist, academic tradition of the Soviet Union was generally lauded, and its success in expanding its scientific manpower considered as exemplary. Many influential essentialist philosophers of the time, including Arthur Bestor and Hyman Rickover, the latter an admiral in the United States Navy, lobbied for the return of academic excellence in public schools. They rejected the child-centered philosophies of the preceding era and urged a strict return to the core disciplines, especially mathematics and science. This was the first "back-to basics" era. It was based on the ideas of these thinkers that the curriculum of schools was reformed.

In 1958, the United States passed the National Defense Act, which was intended to "ensure trained manpower of sufficient quality and quality to meet the defense needs of the United States" and authorized one billion dollars of aid to schools to fund the curriculum in mathematics and science and foreign language education (Brademas, 1987). The federal government was again assuming a national role in education, in collaboration with the states, by legislating a national strategy for defense.

Mathematics was intended to prepare elementary and secondary high-aptitude students. The focus was on students with talent in mathematics for the space age. It was then that the New Mathematics was introduced. This new approach to mathematics, which still exists in a typical eighth-grade textbook, included three new components:

(1) Set theory and notation;
(2) Structural properties of mathematics involving the commutative, associative, distributive properties, whole numbers, integers, rational numbers, additive and multiplicative inverses, and solution of equations; and
(3) The use of bases other than the Hindu-Arabic base ten.

The form of mathematics that was called the "mew math" was really a new approach to mathematics. The problem was that these topics appeared in an abstract manner in textbooks, not connected to any practical application. Moreover there was a gap between set theory and basic computation that mystified parents and teachers. There was also a gap between the subject matter content and methods of teaching the subject (Urban & Wagoner, 2000).

Teachers were unprepared for teaching by the structural approach, and they had insufficient professional help for supporting the change. Resource materials that were sent home mystified both parents who could not help children with homework and of course consequently did not understand the concepts. The result was that new mathematics was discontinued at the lower grade levels. The mathematics curriculum, however, had been transformed and concepts in new mathematics still appears in textbooks (Huetinch & Munshin, 2000). Many Americans questioned the impact of new mathematics on the educational system and on society. Writers such as Morris Kline asserted that the new mathematics was a curriculum suitable for mathematicians. He suggested the need was for a curriculum suitable for the ordinary person who would enter a variety of professions (Kline, 1973). Most of the new mathematics was directed to students with high ability. In an attempt to expose all students to higher mathematics, the new mathematics movement catered to the top students rather than to the marginal ones. The movement was designed to prepare scientists and mathematicians to compete in the global level and accordingly explored such topics as set theory and non-Euclidean geometry that had hitherto been taught at the college level.

It is uncertain whether the new mathematics failed, as is generally acknowledged. Its impact may be evaluated on the performance of students on tests, on the number of students graduating from colleges and universities with degrees in mathematics, on its continuation of topics in the curriculum, and on its impact on technological dominance of the United sates for which its was implemented. The "new" mathematics was apparently unsuccessful at the elementary level as the performance of students at the lower levels declined as evidenced by the scores published by the National Assessment of Education Progress (NEAP) during the sixties and seventies.

Scores of high school students, however, demonstrated that the number and performance of students at the higher end of the ability spectrum increased. In addition, certain topics have continued in the school curriculum, and the success of students of high ability is evidenced by the success of American scientists in putting a man on the moon and establishing hegemony in the field of technology and space exploration.

The impact of new mathematics would also have been felt on the number of students graduating from colleges and universities with degrees in mathematics and related fields. Examination of the statistics from the National Center of Statistics, of the Department of Education is shown in the Table 2-1.

The trend in the graduation rates shown in Table 2-1 suggests that the number of graduates in mathematics increased significantly at the bachelor's and master's levels and again increased during the 1960s and the 1970s. Graduation rates, however, declined during the 1980s. Data from the National Center for Education Statistics (NCES) show that the percentage of mathematics graduates at the bachelor's level was 2.9 in 1959–1960 and then increased to 3.3 in 1961, 3.9 in 1962, and 3.9 in 1963, 4.0 in 1964, 3.9 in 1965, 3.8 in 1966, and 6.0 in 1967. It may be cautiously interpreted that the new approach to mathematics that had been aimed at students with high ability paid off.

Table 2-1. Number and percentage distribution of graduates in mathematics and related science professions

Period	Total math and science	Percentage math and science	Percentage math
	Bachelors and first professional		
1960–61	395,198	28.8	3.3
1970–71	863,000	21.9	3.5
1980–81	1,333,000	19.3	3.3
	Masters		
1960–61	81,735	24.8	2.7
1970–71	224,000	19.7	3.5
1980–81	396,000	17.9	3.4
	Doctorate (except first professional)		
1960–61	10,575	47.9	3.3
1970–71	32,000	45.8	4.6
1980–81	68,700	37.9	4.7

Source: U.S. Department of Education: NCES, 1960-81

Mathematics and Equity and Romantic Humanism in the U.S. during the 1960s

The 1960s were momentous in the history of America. In these years tensions about the Cold War abated, and attention was focused on the domestic scene, which was characterized by the political issues, such as civil strife, urban violence, abject poverty, racial segregation, and race and gender inequity. At the same time ideologies based on humanizing employed compensatory-based strategies. The complementary pedagogy was naturalistic and was based on the philosophies of Rousseau, on relevance curriculum, and on open education. Main thinkers were John Holt, George Denniston, Herbert Kohl, and Ivan Illich, who believed in self-education, and Charles Silberman, who promoted his open classroom learning experiences.

During the 1970s, pedagogy was student-centered and based on the authority of Piaget, whose theory was a developmental approach to education. He advocated cognitive theories with teaching methodologies ranging from the concrete to the abstract. Several summer institutes and teacher education departments trained teachers to give instruction based on the developmental needs of the child. Thus the use of "hands on" methods involving manipulatives for teaching mathematical concepts became widespread in American classrooms. Educators also used open classrooms, based on Silberman's theories, especially in kindergarten classes.

According to Carlson (1996), the 1950s were "awash with innumerable curricular and organizational experiments, many of which left faint imprints on the sandy beaches of curricular reform." He explained that failures included too much emphasis on content and not enough on how to teach content.

Relevance and Back-to-Basics of the 1970s

During the 1970s, the problems experienced by the economy caused a conservative reaction that exhibited itself in the desire for cost saving measures in education. There was a reaction against the open classroom methods that had resulted in unsatisfactory student outcomes. Business and industry expressed dissatisfaction with the skills of high school graduates. The unsatisfactory mathematics results caused a conservative reaction to education. The prevailing ideology became another "back-to-basics" or a return to efficiency and productivity of the early twentieth century. New concepts included accountability performance assessment and competency-based instruction. Achievement of basic skills was seen as a cost-saving device.

Indeed the several commissions during this period emphasized reduction in education provisions, such as narrow curriculum, compulsory school leaving age, and alternative arrangements for obtaining a diploma. Competency-based curricula included a program with a highly specific set of competencies, sequenced according to grade levels. The teacher was directed to teach a competency at the next developmental stage as the student mastered each level through the curriculum. States set requirements for graduation, and students had to attain minimum competency to receive a diploma. The curriculum demanded mostly computational exercises and concepts such as probability. Students learned algebra in the eighth grade rather than starting in the ninth grade.

Despite their reforms, performance scores on achievement tests in mathematics, especially the NAEP, showed significant declines during the 1970s. In 1973 average score on mathematics was 304 for seventeen-year-olds, which declined to 300 in 1978. It was 266 for 13-

year-olds and declined to 264 in 1978 olds. Scores for 9-year-olds remained stable at 219.

Academic Excellence of the 1980s

The 1980s were characterized by vast technological advances that followed a decade of economic difficulties during the 1970s. During that period, the United States experienced an economic transformation regarding international competition, and the interdependent global economy emerged. Critics contended that an academically inadequate curriculum, especially in mathematics and science, had dulled the nation's competitive edge (Gutek, 1965).

According to Gutek (1965) the retrenchment of the seventies soon evolved into a pursuit for excellence, and the back-to-the-subject-centered curriculum, "get tough" approaches of the 1950s. Based on a report called the *Nation at Risk*, the country was exposed to a new set of fears: these were not the old cold was fears, but the fear of economic and technological dominance of Japan. The report dramatized the shortcomings of the school in militant language: "If an unfriendly foreign power attempted to impose on America, the mediocre educational performance that exists today, we might well have viewed it as an act of war." We have, in effect, been committing an act of "unthinking, unilateral educational disarmament." The United States' school system compared adversely with that of Japan on performance in mathematics and science. Experts in education suggested that American schools could profit by imitating the curriculum of Japan. Reformers based their assertions on the progressivism and humanism of the past.

The reforms emanating from the *Nation at Risk* report included discipline-centered education, increased graduation requirements, course work devoted to academic subjects, and emphasis on mathematics and science. "By the eighties the issues were more rigorous academic standards for students and teachers. Forty-two states increased math requirements, and 34 increased science requirements" (Carlson, 1996).

In 1984 Congress passed another law similar to the National Defense Act, authorizing $425 million per year to increase the supply of teachers of mathematics and science in elementary and high schools, and included magnet schools. The goal was excellence in education. In 1980 the National Council of Teachers of Mathematics (NCTM) published its *Agenda for the Eighties,* which in part was a response to the back-to-basics movement of the seventies. The issues were both mathematics content and pedagogy, and it must be noted that performance scores were on the decline during the seventies. Investigation identified the problem. It was agreed that schools taught mostly computational skills geared for passing examinations in standardized tests that were not enough to prepare students to deal with real-world problems that demanded mathematical reasoning and problem-solving situations. It was a fact that students were given word problem-solving activities, but indeed these problems were tied to computation, and the problems were not related to real-life experiences.

In 1989, based on the results of the investigation, NCTM published *Curriculum and Evaluation Standards for School Mathematics*. NCTM proposed that problems should utilize exploration approaches, starting over again when a particular approach was found to be fruitless, consulting with others on strategies, and checking results within the context of the given situation and practical constraints (Curriculum and Evaluation Standards, 1989).

George Polya was a pioneer who proposed a heuristic problem-solving approach. Polya said, "Thus the teacher of mathematics has a great opportunity. If he fills his allocated time with grilling his students in routine operations he kills their interest, and hampers their intellectual development. But if he challenges the curiosity of his students by setting them problems proportionate to their knowledge, and helps them to solve their problems with stimulating questions, he may give them a taste for independent thinking" (Polya, 1985).

The NCTM conceived of mathematics for all students, not just for the most able, as was the case during the period of the new mathematics of the 1960s. It proposed three types of standards: content, process, assessment. Content standards included numeration, estimation, computation, functions, geometry, algebra, statistics, trigonometry, probability, discrete mathematics, calculus, and mathematical structure. The process standards were problem-solving, reasoning, communication, connections, and representation.

These standards were also directed to research-based teaching strategies that are based in cognitive science, especially the inquiry-constructivist approach, and include pedagogy such as meaningful experiences based on real-life problems, cooperative learning, technology, and authentic evaluation standards (Cangelosi, 1996).

National Goals Standards of the 1990s

Discussions of national goals and standards had their genesis in the excellence movement of the 1980s. It began with a meeting of governors in 1989, which set goals for improving academic standards, assisting with school readiness issues, ensuring safety and integrity of the school environment, and culminated in a report of the United States Department of Education in 1990, which also set national standards as Goals 2000 and America 2000. A Congressional Act of 1990 titled "Excellence in Mathematics, Science, and Engineering" was passed to promote excellence in mathematics, science, and engineering programs.

In 1994 Congress created A National Standards Improvement Council to develop national standards and further authorized the NEAP to be the instrument for the evaluation of Goals 2000. By these acts, the federal government was ensuring that mathematics curriculum contained the subjects that would promote economic development and give the United States the edge it needed to compete with Japan and other industrialized countries.

As a result, several professional organizations developed curriculum standards for the subject area. Foremost among these professional organizations was NCTM, which had set content and pedagogical standards for mathematics in 1989. It added two volumes, "Curriculum and Evaluation Standards for School Mathematics, 1991," and "Assessment and Standards for School Mathematics, 1995." These contents were further complied into one document named *Principles and Standards for School Mathematics, 2000*, which set the agenda for the new millennium.

In addition to national standards, most states have promulgated their own standards in mathematics. The standards in New York represent an integrative approach and are called *Mathematics, Science, and Technology.*

Incorporating Technology

The inclusion of technology in the curriculum is an important component that has been incorporated since the 1990s. Heutinch and Munshin (2000) stated that "technology makes an additional topic in mathematics less important, others more important, and new topics possible." Examples are discrete mathematics; transformation geometry; connecting representations of functions such as graphs, data tables, and equations; and more accurate representation of three-dimensional graphs. Use of three-dimensional calculators, access to the Internet, and spreadsheets are now commonplace in the school curriculum.

Evaluation of NCTM Standards and Methodology

Workers' skills in mathematics and science are a critical component for economic and military competitiveness in this global technological age. The mathematics curricula prepare students for workplace requirements and educate professionals such as engineers. The performance of students on tests in mathematics thus is an indication of how successful they will be as workers, and hence predicts the quality of the global economy. Accordingly the NCTM standards in mathematics and science programs are the measure for evaluating school programs, and the performance of students on these standards-based examinations indicate the success of the NCTM standards.

The United States Department of Education is responsible for providing a picture of the state of educational achievement in the nation's schools. It has established the National Center for Educational Assessment (NCES) to collect criterion-referenced measures of achievement in all subject areas including mathematics and science. The tests, which are congressionally mandated, are known collectively as the National Assessment of Educational Progress (NAEP), the results of which are published as the nation's report card of educational progress.

The tests track longitudinal achievement on sets of knowledge and skills by state and the results are reported as a composite score. NEAP are criterion-referenced tests based on national standards established by the NCTM. Content areas include number sense, properties and operations; measurement; geometry and spatial sense; data analysis, statistics and probability; and algebra and functions.

The Nation's Report card of 1999 published by the NEAP stated, "Generally the trends in mathematics and science are characterized by declines in the 1970s, followed by increases during the 1980s and 1990s, and mostly stable performance since then." For example, average mathematics scores, which had been 304 for seventeen years in 1973 declined to 298 in 1982 but resurged to 308 in 1999.

According to the Condition of Education (2000), NEAP information shows that "Mathematics assessment, that reflects recent emphasis on mathematics standards developed by the NCTM, showed that scores for the 4th, 8th, and 12th graders increased between 1990 and 1996. The statistics showed that there was statistically significant improvement in all areas. In 1996 the average score for fourth graders was 224; for eighth graders, 272; and for twelfth graders, 304. Each of these scores was an improvement over 1990 and 1992. The standard "moderately complex procedures and reasoning " showed an increase from 52 % in 1978 to 60% in 1996; and the scores on the standard problem solving showed a slight increase from 6% in 1982 to 7 % in 1996." (Condition of Education, 2000).

There was similar improvement in student performance on science examinations. In essence, because the NCTM standards have been implemented in schools' curricula, the performance of students on mathematics examinations improved between 1992 and 1996, indicating that the standards have been successful.

Conclusion

Historical, economic, and sociopolitical factors have shaped mathematics curriculum and pedagogy, which reflect social realities, government policies and agencies, and professional organizations. Mathematics curricula has been translated into standards for content, process, and assessment. These standards have been evaluated objectively for their effectiveness, that is, the performance of students on criterion-referenced tests. In socioeconomic terms, the total numbers of skilled workers in mathematics and engineers have increased.

ASSIGNMENT

Based on your reading of this chapter and of the references that follow, answer the following questions.

1. Evaluate the performance of students on the NAEP tests in mathematics in all fifty States of the Union.

2. Prepare a graph of the number of graduates in mathematics since 1950.

3. Outline the major reforms in mathematics curriculum since 1950.

4. Discuss standard-based mathematics as a strategy for reform.

5. Discuss the role of professional organizations and the States in the reform of mathematics.

6. Describe the involvement of the States in standard-based mathematics and the impact on the performance of students.

7. Write a comparative description of traditional and of standard-based mathematics and the impact on pedagogy.

Figure 2-1. Miraflores Lock, Panama Canal

Chapter 3

How Mathematics Shaped Two Societies: The United States and Russia

Mathematics education has the potential to shape society. An educational system prepares its youth for the type of society envisioned by economic, political, and other leaders, mainly through the control and direction of the school curriculum. Modern industrialized societies including the United States and the Soviet Union with their ideas of economic superiority, space exploration, and cyberspace utilization, have tended toward setting socioeconomic goals that are based on the architectural need for scientific and mathematical knowledge.

This chapter discusses the relative impacts of employing the strategy of mathematics and science education by the countries under review, taking into consideration their differing traditional and democratic political systems. It is a comparison of two distinct entities and approaches. The first is a state-controlled system of education, founded in 1783 by Catherine II, Tsar of Russia, as a free school, but reshaped by Tsar Nicholas I in 1829 along class lines. The second is the free-choice Western American education system that had fledgling beginnings in the seventeenth century but whose genesis as a public institution occurred in the later nineteenth century. Johnson (1950) has pointed out that the Soviet school system has been largely dominated by "Tsarist, elitist, and solidly conservative academic tradition." By contrast western education has its basis in democratic and progressive ideologies (Gutek, 1995). Both countries, however, have used the education system for the similar goal of national development. The emphasis of this discussion is on curricula, particularly in mathematics, that are designed to further the respective economic, political, and social goals. Modern nations, including the United States and Russia, have demonstrated how mathematics curricula have helped to create dynamic industrialized societies.

The chapter investigates two periods of curricular reform in mathematics: the era of the 1960s when the race for superiority in space between the United States and the Soviet Union began, and the later 1990s when Russia again became dominant and replaced the body politic of Soviet Union. During the earlier period, the respective governments enacted legislative policies for the implementation of policies and programs that intended to alter profoundly the technological balance of power: the National Defense Act of 1958 in the United States and the Khrushchev School Reform of the same period in the Soviet Union. Outstanding reforms were made in the content and pedagogy of mathematics and science education. This chapter examines the policies of both countries and analyzes their relative impacts in the contexts of the demographics, economies, and histories of the countries. The effectiveness of the policies is assumed to be determined by the output of manpower with skills in mathematics and science and related engineering and computer technologies. Measurements and interpretations of the numerical output from universities of skilled personnel per thousand of the respective populations are based on secondary data.

The second period of reform was the 1980s and 1990s, when there was a new set of economic and political preoccupations of the United States and the Soviet Union, later Russia. This chapter examines the contexts and goals of the two countries and the outcomes of the implementation of the standard-based mathematics curricula and testing programs in that period so as to determine the effectiveness of mathematics education as a strategy of national development. Secondary data provide the basis of measurement of the performance of students on the International Assessment of Education Progress (IAEP) in mathematics and science and enumerate the supply of technocrats and professional manpower created by the application of mathematics and science curricula.

American Mathematics Education During the 1960s

In the United States, education is generally the responsibility of the fifty states individually. Since the Northwest Ordinance of 1787, however, the United States federal government has played a prominent leadership role in education reform when problems such as national development, welfare, and defense have become critical. Unlike the Soviet Union, the United States did not initially have a tradition of technical education. According to Hans (1964), technical education began in the nineteenth century, when it was a utilitarian preoccupation of the average citizen who became very adept in creating and managing technical equipment. The federal government had little role in technical education, which accordingly grew haphazardly between 1860 and 1890. Indeed technical education was highlighted by a series of international exhibits where Moscow stole the show. U.S. federal government involvement in engineering education began with the First Morrill Act of 1862, which made resources available for the development of agricultural and mechanical arts, followed by the Smith-Hughes Act of 1917, for vocational education. Hans (1964) stated that development of technical education was precipitated by the emerging industrialization of the nineteenth century, which realized revolutionary proportions in the twentieth century.

When the Soviet Union launched an earth satellite into orbit in 1957, Americans were alarmed to find that their arch-competitor possessed a distinctive technological edge. According to Carlson (1996), "No single event has had greater impact on public opinion or more strongly reinforced the conclusion that the American school system was failing in the race with the Soviet Union for world leadership." Politicians blamed the school system for failing to develop technical education. Educators blamed the "soft pedagogy" of the previous progressive era and many influential essentialist philosophers of the time and lobbied for a return of academic excellence in public schools. This was the first "back-to-basics era."

The National Defense Act of the United States

Congress passed the National Defense Act in 1958. Brademas (1987) stated that it was the beginning of the space race that was to alter profoundly the relationship of the U.S. Federal Government to the nation's schools and universities and their curricula. The goal of the National Defense Act was to "ensure trained manpower of sufficient quantity and quality to meet the defense needs of the United States." Congress authorized aid to fund research in universities and implementation of mathematics, science, and foreign languages in the curriculum of schools. The focus was on students with talent in mathematics to develop skills commensurate with the space age. It was then that the New Mathematics was introduced. New Mathematics included set theory and notation, structural properties of mathematics, the use of bases other than the Hindu-Arabic base ten, and non-Euclidian geometry that had previously been taught at the college level. Critics have contended that the New Mathematics failed. So it did if one only considers its impact on elementary schools and the average student. In fact the New Mathematics was directed at students of high ability, and for these students, it did not fail. The impact would be felt on the number of students who graduated from universities with degrees in mathematics during the 1960s and 1970s. Statistics from the National Center of Statistics, U.S. Department of Education, show that between 1959 and 1966 the graduation rates of students in mathematics as a percentage of the total output from all universities in the United States increased.

The trend in the graduation rates seemed to suggest that the number of graduates in mathematics increased significantly at the bachelor's and master's levels during the 1960s and the 1970s and again declined during the 1980s. At the bachelor's level, the graduation rates in mathematics almost doubled between 1959 and 1969, from percentages of 2.9 in 1959, 3.9 in 1961, 4.0 in 1963, and 3.9 in 1966. It may be cautiously interpreted that the new approach to mathematics that had been interpreted as a failure actually paid off by increasing the pool of manpower with mathematical-related skills. Further, certain topics have been retained in the curriculum.

Soviet Mathematics During the 1960s

According to Bereday (1964), historically, the Soviet system has had a tradition of taking its youth wherever it wanted them to go, by following tradition, subject matter (particularly in mathematics and science), and the promise of careers. Soviet education has been challenged by "ancient traditions, the military needs of the state, and the growing industrialization." Greek culture, including the Aristotelian tradition, gave academic impetus to Russian education. The Kiev schools established in 1649 were based on Greek education, training in cartography laid the foundation for Russia as a military power, apothecaries precipitated medical training in the seventeenth century, and the efforts of Tsar Peter the Great and his admiration for Western civilization, laid the foundation for engineering and the creation of the Russian navy during the seventeenth and eighteenth centuries. When mining appeared in the eighteenth century, industrialization became the main focus of the Russian economy.

Because the Russian economy had emphasized technology, the teaching of science and mathematics occupied the chief place in the curriculum (Bereday, 1964). At the time of the October Revolution in 1917, there was already a tradition of mathematics, evidenced especially by the works of P. L. Cebysev, the founder of the St. Petersburg School of Mathematics, and A. A. Markov. These mathematicians made sterling contributions to the discipline in statistics, physics, genetics, and economics (Struik, 1987), and the productivity of schools was reflected in the manpower produced. Several renowned technical schools, such as the school of navigation in 1701 and the Academy of Sciences in 1725, were established. According to Bereday (1964), the founding of the Institute of Engineers in St. Petersburg in 1806 inaugurated the growth of technical education that attained international flavor with a series of technical displays.

Timoshenko (1959) deteremined that engineering schools contributed vastly to Russian scientific education and Cremin (1961) suggested that at Philadelphia Centennial Exposition in 1896 Russia literally stole the show and demonstrated its technical superiority. Sir Eric Ashby suggested that it was Russian's prerevolutionary tradition ythat gave direction to that country's scientific and technological strength.

With the revolution of 1917, Russia was absorbed by the Soviet Union. According to Gutek (1995), Soviet leader Nikolai Lenin saw education as a necessary instrument for achieving the economic modernization of the Soviet Union. The Soviet government introduced several policies for the promotion of science and mathematics education for providing manpower resources for economic development. To achieve the goal of creating a Soviet culture, Lenin appointed Anatoli Lunacharski, who introduced many measures especially in pedagogy adapted from western theorists such as Montessori and Dewey.

These policies were discontinued when Joseph Stalin gained power. Under Stalin, the Soviet Union restored formal training in sciences and mathematics.

In 1928 Stalin created a Five-Year Plan based on the idea of the Polytechnic. According to Mallinson (1964), the Polytechnic linked instruction with socially practical labor and training so as to convey greater skill and knowledge to those engaged in industry. The Soviet Union placed more emphasis on technical education than any other country in the world.

In an address to the Komosomol in 1958, Khrushchev outlined his Plan titled, "On Strengthening the Ties of Schools with Life." This was followed by an article in *Pravda* on 21 September 1958, and an address at Rangoon University at which he touted the technological strength of the Soviet Union. There was a return to the Polytechnic on the 1920s and the creation of special schools to train the brightest students in mathematics and science.

The Universities also played a role in promoting pure and applied mathematics, including the work of Luzin, who demonstrated the connection between mathematical theory and application to physics and industry.

Soviet mathematicians made some gains in engineering and productivity, creating a pool of mathematicians who contributed to its technological strength in the fifties and sixties. DeWitt (1962) reported that "during the last two and a half decade the Soviet Union has made enormous strides towards building up its specialized manpower resources. And as a result it had reached a position of close equivalence with or even slight numerical superiority over the United States as far as supply of trained manpower is concerned."

Comparative Effectiveness of United States and Russian Policies of the 1960s

This section uses the manpower produced during the 1960s to compare the effectiveness of the respective policies of the United States and Soviet Union. These national policies, the National Defense Act of the United States and the Khrushchev School Reform of the Soviet Union, were designed to promote mathematics and science education. The outcomes may be evaluated by examining the output of trained manpower with skills in mathematics and science. Bereday (1964) suggested that comparisons of manpower potential must take into account the relative sizes of the territories and their populations in computing rates thousand. Conclusions must also consider the nature of the societies, their histories, economics, and political circumstances. The relevant period selected from World Data (1993) indicates that in 1979, the USSR had a potential pool of 263.4 million inhabitants, compared with the U.S. population of 228.5 million. Also whereas the USSR was the largest country worldwide with land space of 8,649,490 square miles and wasteland of 37.5 percent, the U.S has 3,615,122 square miles with 22 percent wasteland. Historically the Soviet Union has had a tradition of academic technical education, with limited technical application; until the twentieth century, the U.S. had been the reverse. U.S. education and practice have been utilitarian, which Bereday (1964) described as raising a nation at home with skills in automation that astounded the Russian soldiers during the war. By contrast the Russian tractor seemed very technical to the average U.S. soldier. Hence comparisons must be viewed with caution.

A study by Nicholas DeWitt, Education and Professional Manpower (1962) supplied quantitative data on the manpower produced by the two countries. A cursory review of the raw data below (see Table 3-1) portrays the manpower as evenly balanced, with the Soviet Union having a slight superiority according to DeWitt.

Comparisons become more questionable, however, when other factors are considered. Havighurst (1958) contended that the number of scientists and engineers produced has to be evaluated in terms of the total number produced and the total population. He showed that in 1950 twenty-six per thousand engineers were produced in the U.S., compared with the Soviet Union's eleven per thousand. In 1958, these figures increased to twenty per thousand in the U.S. compared with seventeen per thousand in the Soviet Union. The 1962 data above show that in the production rate of engineers the U.S was twenty-three per ten thousand compared to the Soviet Union's nineteen per ten thousand. The indications are that the U.S outproduced the Soviet Union in the number of engineers but the Soviet Union had more doctors per ten thousand of the population.

Table 3-1. Education and Professional Manpower in the U.S. and the Soviet Union

	U.S.	Soviet Union
College Teachers	210,000	90,000
Dentists	90,000	20,000
Engineers	530,000	500,000
Physicians	195,000	280,000
Teachers	1,400,000	2,000,000

Source: *Soviet Professional Manpower*, NSF by Nicholas DeWitt, 1962

The Intervening Years: Soviet Conservatism and Equity and Romantic Humanism of the U.S.

In the years following the National Defense Act and the Khrushchev School Reform, the Soviet Union remained conservative while the U.S. underwent a revolution of civil liberties that was to be used as a point of reference by the Soviet Union and later Russia. The years of the 1960s were momentous in the history of America. In these years tensions about the Cold War abated, and attention focused on the domestic scene and on the Civil Rights issues of social justice and gender equity.

At the same time ideologies were humanizing and naturalistic based on the student-centered pedagogy of Rousseau and of Silberman with his open-classroom learning experiences. According to Carlson (1996) the 1960s were awash with curricular and organizational reforms, many of which failed because of too little emphasis on content. The achievements of students declined significantly. Although the number of pure mathematics graduates from all U.S. universities declined during the 1970s, the graduates in related engineering and computer technologies increased dramatically. The Condition of Education (2000) gives the decrease in the percentages of graduates in pure mathematics as falling 3.0 in 1971 to 1.8 in 1976. The percentage of students graduating with engineering degrees, however, increased from 5.3 in 1971 to 4.1 in 1976 and to 6.8 in 1981, indicating the beginning of a change from pure theory to applied mathematics orientation, which was to continue during the 1990s.

Academic Excellence and Standards of the 1980s and 1990s in the United States

The 1980s were characterized by vast technological advances that followed the previous decade of economic difficulties. During the seventies, the United States experienced an economic transformation regarding international competition and the emergence of the interdependent global economy. Critics contended that an academically inadequate curriculum, especially in mathematics and science, had dulled the nation's competitive edge (Gutek, 1965). The retrenchment of the 1970s soon evolved into a pursuit of excellence, the back-to-the-subject-centered curriculum "get tough" approaches of the 1960s. Based on a report called the *Nation at Risk*, the country was exposed to a new set of fears, not the old cold war fears, but the fear of economic and technological dominance of Japan. *Nation At Risk* dramatized the shortcomings of the curriculum in militant language.

The United States' school system compared adversely with that of Japan on performance in mathematics and science. Suggestions were made that American schools could profit by imitating the curriculum of Japan (Gutek, 1965). Concerns about declining standards of the 1970s led to discussions of national goals and standards and the birth of the excellence movement of the eighties and nineties. Based on this goal, Congress enacted a number of policies to increase the pool of manpower with skills in mathematics and science. Congress

—authorized $425 million a year to increase the supply of mathematics teachers in high schools.

—accepted the Goal 2000 and America 2000 Reports, which articulated the goal of excellence.

—passed an Excellence in Mathematics, Science, and Engineering Act and created a National Standards Improvement Council.

—empowered NCES to conduct ongoing evaluation of the national standards using a series of tests collectively called the National Assessment of Educational Progress (NAEP).

Major players in these reforms were professional organizations that developed standards for subject areas. Foremost among them was NCTM, which set content, process, and pedagogical standards in 1989, 1991, 1995, and 2000.

Following the lead of national standards, states also set their own standards during the 1990s. In New York State established standards based on an integrated mathematics, science, and technology curriculum. In essence, these standards set the agenda for the millennium. According to the Condition of Education (2000), the NCTM standards in mathematics are the benchmark for evaluating school programs. The performance of students on these standard-based examinations indicates the success of the NCTM standards. These standards are measured by the NAEP, which are criterion-referenced tests that track achievement of students, and are reported as composite scores. Performance of students in mathematics on the NAEP improved, indicating that the standards have been successful.

Despite the excellence in the performance of students in mathematics that emanated from the standard-based mathematics curriculum, the percentage of graduates from universities did not increase. Indeed statistics from the Condition of Education (2000) show that the percentage of students graduating from universities with degrees in mathematics decreased from 3.0 in 1971 to 1.2 in 1995. Students, however, continued to be diverted from pursuing careers in pure mathematics and instead studied engineering and computer technologies, with the percentages fluctuating between 6.8 in 1981, 7.7 in 1986, 5.6 in 1991, 5.4 in 1992, and 5.4 in 1995. In effect the potential pool of manpower with mathematics-related skills increased during the 1980s and 1990s.

Humanistic and Standard-based Education in Russia: 1990s

After its dramatic successes of the 1960s, the Soviet Union remained conservative during the 1970s. In 1985 it was ready to change policies. Democratic reforms began in the Soviet Union under Mikhail

Gorbachev in 1985 in what has been called restructuring or *perestroika.* These reforms included calls for democratization, separation of church and state, secular education, municipal institutions, public participation instead of centralized authority, and the content of education to meet the needs of society. Nikandrov (2000) claimed that these ideas had existed for decades in Russia in embryonic form and that many had been implemented under the *Statute of the Unified Oriented School* by the first Commissioner of Education, A. Luamnacharsky after the Revolution in 1917. These had ceased after the 1920s and were again initiated under Gorbachev.

In 1991 the body politic of the Soviet Union was replaced by a number of independent states, chief of which was Russia, which retained its status as the largest country in the world. The law of 1992 was amended in 1996 to state the principles of an education policy proclaiming a humanistic nature of education (Nikandrov, 2000). Reforms included free access to education for all, adaptation of education to the needs of students, freedom, and pluralistic education. Present humanistic ideologies are developmental in focus, based on the philosophies of Vassily Davydov, who was inspired by the Russian psychologist, Lev Vygotsky.

Accompanying the standards was a testing program to assess the achievement of students in mathematics. This testing program, which is similar to the NAEP, has already been implemented in Russia, but like the NAEP, it is not mandatory. The 1996 reforms also proposed parental choice in educational institutions, similar to the U.S. agenda.

Current education in Russia is paradoxical and may be described as "decade skipping." By the law of 1996, it appears that Russia is at once adopting a backward humanistic policy that was tried in the United States during the 1970s and is also adopting a modern standards-based policy that currently exists in the United States. Generally Russian education now has less specific goals than in the era of the Soviet Union. By introducing these reforms, Russia appears to be abandoning its tradition of quality education for the selected group of students to train them to meet the needs of the economy and is adopting a policy of equality and humanistic education that was initiated in the United States in the 1970s. Note that these policies did not promote academic excellence in mathematics education in the United States. Russia might well exercise caution in its humanistic education policies as far as mathematics is concerned.

In humanistic education, all students are to be exposed to mathematics equally instead of selected students. However, these policies are more idealistic than real. The policies are not yet regulations, as under a state-controlled government, but are rather goals. Further, observation of mathematics classes at a gymnasium and high schools in St. Petersburg revealed that there is still excellence in the content of mathematics in Russian schools. Beginning in grade six, all students take three classes in algebra and two in geometry per week throughout their school careers. There are also free-choice schools for gifted students who will pursue further education in mathematics at the University.

Comparative Productivity of Schools and Professional Manpower: United States and Russia, 1990s

Both the United States and Russia have introduced standards-based mathematics and science curriculum as well as methods for assessing the effectiveness of the curriculum. The International Assessment of Educational Progress (IAEP) is the recognized instrument for comparing the performance of students in countries around the world. The Condition of Education (2000) stated that data from the second international test administered in 1991 to thirteen-year-olds in fifteen Republics of the Soviet Union and to 20 other countries of the world showed that Soviet students had better composite average scores in both mathematics and science than U.S. students. In 1999, when the Soviet Union was succeeded by Russia, the performance was maintained. Data from the *Statistical Almanac* showed that on the IAEP administered to eighth graders in thirty-eight countries, the average scores for U.S. students were 502 for mathematics and 515 for science; for Russia the comparative scores were 529 for mathematics and 526 for science.

Comparisons must take into account, however, the structures of the school systems. The school systems in the Soviet Union and Russia are selective; that is, the gifted are chosen for careers. By contrast the education system in the United States is based on the concept of the "common school"; that is, schools that are open to all students equally. Hence the outcomes represent different processes and must be viewed with caution. Data on the output of manpower with skills in engineering and science have been described in the *World Statistical Abstract* (Mutt, 2000). The 1994–1995 data show that the United States had less landscape than Russia but a larger population, greater resources, and seemingly greater output of manpower. The Gross Domestic Product (GDP) per capita for the United States was $27,500 compared to $5,300 for Russia; expenditure on education was 5.5 percent of GDP in the U.S. compared to 4.4 percent in Russia. Expenditure on research and development was $171 trillion dollars for the United States compared to about $1.3 trillion rubles for Russia. Based on these data, manpower output for scientists and engineers amounted to thirty-seven per ten thousand of the population for the United States compared to forty-three per thousand of the population for Russia. The comparative figures for doctors, per ten thousand of the population, were 420.6 for the U.S. compared to 221.7 for Russia. The data show that Russia, where students' performance on the IAEP tests in mathematics and science was better than that of the U.S., the manpower productivity in engineering and science was slightly better. The U.S., however, significantly outperformed Russia in the production of manpower in the medical field (see Table 3-2).

Table 3-2. Comparative Demographic Data, U.S. and Russia, 1994–1995

	United States	Russia
Land area (sq km) or	9,166,600	16,995,800
(sq miles)	3,540,321	6,592,850
Population	263,034,000	148,291,000
Population density*	74	22
% expenditure on education per GDP.	5.5	4.4
Engineers and scientists	962,700	644,000
Physicians per population	420.60	221.7

Persons per sq km.
Source: Statistical Abstract of the World (Mutt, 2000)

Provision of Resources

Another important provision of the 1992 and 1996 Russian laws on education was that "No less than 10 percent of national income should be allocated to education" (Nikandrov, 2000). This was a goal rather than a reality. In 1994–1995, Russia spent 4.4 percent of its GDP on education and the U.S. federal government actually allocated

5.5 percent of GDP to all levels of education. Again this has to be interpreted cautiously because of the sizes of the gross budgets and the comparative populations of the two countries.

Education in the U.S. also is funded at the federal, state, and local levels. Generally only seven percent of the total expenditure is contributed by the federal government. The fact that a sizable proportion of the budget is allocated to education nevertheless indicates that mathematics education is given priority as a national goal in both the United States and Russia.

Conclusion: Mathematics for the Workplace and the Global Economy

In the latter half of the twentieth century, the U.S. economy has continued its dynamic economic growth while the centralized Soviet Union has been redesigned into a number of independent states. The education system serving the political objectives of the two societies has been implemented successfully, demonstrating that education is an independent variable that may be used successfully despite the political systems in which policies are similar and in some ways programs are transferable. Moreover, whether the base is academic or practical, the ends are similar, and mathematics education plays a critical part in the development of technology. Further, the establishment of the international space station demonstrates that the technology is at its finest when there is cooperation between the two countries. There is still the belief that mathematics, which itself is an international language of communication, may be the dynamic instrument and central common engine that may further international cooperation and that the education systems may shape the societies by producing the required manpower.

Education has dual objectives in America and Russia: preparation of students for the economy and for individual development. Education is an investment; manpower with skills in mathematics will shape society. Mathematics, science, and technology will no doubt be critical, creative, and dynamic independent factors in manpower development . Comparison between the educational systems U.S. and Russian has demonstrated that mathematics is the engine that will fuel the global economy. In a world shaped by mathematics, science, and technology, the strength of nations might be determined more by cooperation than by competition, and this approach may augur well for the future.

ASSIGNMENT

Read the references that follow and consult data from the National Center for Education Statistics and the World Almanac to answer the following questions.

1. Describe in detail the historical tradition of Russian mathematics.

2. Compare the effectiveness of the traditional Russian mathematics and the new mathematics of the United States in providing manpower for space superiority.

3. Compare the current reforms of mathematics in the United States and Russia.

4. How effective have been these current reforms in providing manpower with skills in mathematics and science

5. Evaluate humanistic mathematics in the United States. Should the Russians copy this model?

6. Write a description of the structure and mathematics curriculum in the United States and Russia.

Figure 3-1. Photograph of a Monument to Czar Nicholas 1, St. Petersburg, Russia

Chapter 4

Education in Japan and the United States

Following World War II, Japanese society was reshaped along western lines, and as a concomitant, the education system was restructured to promote the economic and social development of Japan. From inauspicious beginnings, Japanese education is now modern and productive, and its new challenges are issues concerning globalization and internationalism.

The most significant characteristic of Japanese society is its homogeneity. Of a population of 125 million in 125 islands of mostly city dwellers, there is only one percent minorities. Tokyo, the capital city with a population of twenty-eight million inhabitants, is the largest city in the world. The development of modern Japan began after World War II with the defeat of Japan. During that time Japan became heavily industrialized, and the country placed a high value on education as the catalyst to employment and a better lifestyle.

Goals of Japanese Education

The modern Japanese education system began with the promulgation of "The Fundamental Law of 1947." Kanaya (1995) stated that the objectives of public education in Japan are

1. The development of broad-minded healthy individuals.
2. The development of the spirit of freedom, self reliance, and public awareness.
3. Preparation of a Japanese citizen to live in a global society.

Japan has a national, centralized educational system. Responsibility for the direction, control and governance, and financing of the educational system lies with the Ministry of Education, Science, and Culture. The Ministry sets the curriculum framework, including the national curriculum objectives, content, and standards for time allocation.

The work of the Ministry is decentralized. The municipalities and individual schools participate in the management. Municipalities are governed by a five-member board that implements the guidelines that have been set by the Ministry. From the guidelines, individual schools develop their own courses for instruction. Responsibility for finance is shared by the national government, the municipality, and the individual school.

The Structure of the Educational System

The stages of the Japanese educational system parallel those of the United States. Despite the similarity of structure, there are fundamental differences. The Japanese system is selective, with access to second-stage high school based on merit as opposed to the American school system, which is open and common to all students. In addition where Japanese education system is a national system, responsibility for education in America devolves to the states.

The Japanese educational system is a three-tiered system from elementary to university levels. After completing the lower elementary school, students take an admission examination to gain access to the upper secondary school. Goya (1994) described this as an elimination examination, as typically only a small percentage of the students are eliminated. For example, of 304 students taking the examination, 300 are admitted to the upper secondary school. Kanaya (1995) described the stages of the education system thus:

Elementary School:	Students begin attending school at the age of six and attend elementary school, from grades one to six. The pupil teacher ratio is 20:1.

Lower Secondary School:	The second stage is lower elementary school for three years. Here the student teacher ratio is 17:1 and classes are of 50 minutes' duration. At the end of lower elementary school students take an entrance examination for access to the upper elementary school.
Upper Secondary School:	At the upper elementary level, seventy-five percent of the students are in academic courses and the rest are in vocational tracks such as technology, foreign languages, and computers.
Special Needs Schools:	Since 1988 students with special needs attend secondary schools and obtain diplomas by a credit system rather than by examinations.
University:	Thirty-four percent of Japanese secondary school students attend university based on a competitive examination.
Examinations:	"Unlike the United States, there is no external examination, no public or private testing service that scores and distributes, and scores a set of examinations common to students across the country" (McNergney & Herbert, 1995). The school system is responsible for assessing, determining, and promoting students. The individual school determines completion and awards certification.
Extra Classes:	Students wear uniforms to schools. Many students attend a private school after school classes. They are offered two types of courses. The first type is enrichment classes in areas such as music, arts, and physical education (*obeika-goto*). These constitute lifelong education. The second type of after-school classes are supplementary, for improving academic performance (*juks*). Their purpose is to provide remedial and enrichment education to help students meet the demands of the curriculum.

Operations: Generally Japanese students attend school more days per year than American students: 240 vs. 180 in American schools (Stevenson & Stigler, 1992). In Japan a typical school year includes sixty-five to seventy afternoons of nonacademic activities. Three or four days per year are devoted to cleaning (Goya, 1993) and Japanese students enjoy longer lunch periods.

Japanese Pedagogy

There are two perspectives on the methods of pedagogy in Japanese schools. Desmond (1995) suggested that there is much rote learning and conformity. On the other hand, some researchers have reported that the system is not authoritarian; teachers act as facilitators. Also Stevenson and Stigler (1992), who observed fifth grade classes, described Japanese methodology as conforming with the approach known in the United States as constructivism.

Comparison of U.S. and Japanese Pedagogy

The excellence movement of the 1980 and 1990s was intended to improve the competitive edge of the United States over Japan. In 1984 and 1990 Congress allotted additional funds to improve performance in mathematics and science. It also dedicated funds to the establishment of structures such as the National Assessment of Educational Progress (NAEP) to evaluate educational progress. The provisions certainly helped in the improvement of mathematics and science curricula and contributed to the improved performance of students on national and international tests.

However, U.S. students did not outperform Japanese students on international tests in mathematics and science. These tests were administered to eighth graders and were called the *Third International Test of Mathematics and Science Study* (TIMSS). The National Center for Education Statistics (NCES), which administered the tests, also analyzed the mathematics scores, the curriculum, and the textbooks used in schools in both the United States and Japan. NCES determined that these resources were equivalent.

The TIMSS report (see Table 4-1) stated that eighth grade Japanese students scored significantly higher than U.S. students (Huetinch & Munshin, 2000). The U.S. Department of Education showed the scores in 1998 as 497 for the United States and 600 for Japan. (U.S. Department of Education, 1998).

Huetinch also commented that although the topics were similar, there was a difference in the pedagogical methods. In the U.S., specific topics are covered for considerably longer, for eight years on average, compared to three years in Japan, and U.S. students cover many more topics each year than the Japanese. The comparison continued with analysis of instruction in through fifty to one hundred Japanese and U.S. eighth grade classrooms. TIMSS summarized U.S instruction as emphasizing familiarity with many topics rather than concentrated attention, which seems to foster an emphasis on quantity rather than quality.

NCES, which administered the international tests, compared the performance of students in the United States and Japan. The report concluded,

> When the videos are viewed, Japanese teachers, in certain respects, come closer to implementing the spirit of current ideas advanced by U.S. reformers than do U.S. teachers. The content of U.S. mathematics classes requires less high level thought than classes in Germany and Japan.
>
> U.S. mathematics teachers' typical goal is to teach students how to do something, while Japanese teachers' goal is to help them understand mathematical concepts.
>
> Although most U.S. mathematics teachers report familiarity with reform recommendations relatively few apply the key points in their classrooms.

Perhaps the most significant difference is the structure of the two school systems. Japan, like most countries of the world, has an elitist system that "creams off" the brightest students and trains them in selective schools. By Contrast, the American education system was founded on the concept of the "common school," which at least in theory affords everyone equality of opportunity. For example, the TIMSS report shows that of the forty-one countries in the sample, "The average eighth-grade U.S. lesson deals with mathematics which is comparable to the ninth grade level by Japanese system, and the seventh grade by international standard."

Table 4-1. Comparison of U.S. and Japanese Mathematics Based on the Third International Mathematics and Science Test (TIMSS)

United States	Japan
A. Content	
College professors evaluated content quality, with the following indicators from high to low. Most mathematics lessons included a mixture of concepts and applications. The difference was in the method of delivering the concepts.	
1. Concept Development	
1/5 of U.S. teachers practice the the conceptual approach.	3/4 of Japanese teachers develop concepts
0% U.S lessons include proofs proofs	53 % of Japanese teachers use proofs
2. Reasoning	
Quality of mathematics content is evaluated by coherence of math concept across the lesson, and degree of deductive reasoning. Rating was high to low.	
0% received high rating	39% received high rating
89% received low rating	11% received low rating
3. Practicing Routine Procedures	
90%	41%
4. Applying New Concepts and Inventing Solutions	
Not known %	44%
B. METHODOLOGY	
1. Structure of Lesson	
Goal is problem-solving	Goal is understanding; problem- and problem-solving is the means
2. Instructional Processes	
Instruction has 3 steps: Statement of rule Explanations using examples Solving problems	Order of activity is reversed: Presentation of problem Reflection and analysis Solutions emerge
3. Coherence of Lesson	
Lessons have many topics, much interruption	Links or connections are made between parts of the same lesson

Source: Third International Mathematics and Science Test, 1999.

The Japanese selective system caters to the brightest students, whereas the U.S. system deals with the average student. The report suggests that a student at a ninth grade level of achievement in Japan is being compared with a student at eighth-grade level of achievement in the U.S. The comparisons are unequal. As in the case of Russia, comparison between the two systems is difficult at best.

Despite the fact that Japanese children outperformed the U.S. children on tests in mathematics and science, the same cannot be said of the economy. Note that the TIMSS report described the type of work done in both school systems as "practicing routine procedures and applying concepts to novel situations and inventing solutions."

Although ninety percent of U.S. students spent time in practicing routine solutions compared with forty-four percent of Japanese students, forty-four percent of Japanese students spent time inventing new solutions. Apparently inventing solutions is not related to the creativity of workers in the economy. Perhaps the U.S. workers could do even better if it made a change in the methodology of teaching mathematics and practiced making inventions as part of the mathematics curriculum. The National Council of Teachers of Mathematics (NCTM) provides standards for improving the teaching of mathematics. The question is, "Are they being implemented in schools?

Despite the fact that Japanese students outperformed the U.S. on mathematics tests, the same cannot be said of the economy. This might be related to the practice of performing routine work and inventions in mathematics. It might also be related to the utilitarian principles on which the U.S. established its economy. Comparison has to take into account economics, history, politics, and technological achievements.

Preparing Students for the Workplace and the Economy

American education, like education in other countries, was established to prepare students to participate in the economy. This began with the fledgling industrialization of America, when Horace Mann organized education for the new industrial age. In the third millennium, the economy is a global one, which has impact across national boundaries. The economy increasingly involves technology. Mathematics, science, and technology will be the answers to making this economy work in the future. As in the case of Russia, comparison between the Japanese and American school systems is difficult.

Conclusion

The education systems in the United States and Japan have great similarities in terms of the curriculum and requirements. The structure of the school systems differ however, from the common open system of the United States to the selective system of Japan. Pedagogy in Japan is more like the methods advocated by American educators, that is, the constructivist approach. Performance of students on international tests is significantly better in Japan than in the United States according to reports on the international tests. Mathematics is regarded as the engine for providing workers for the economy in both the United States and Japan. Mathematics produces the skilled manpower in mathematics and engineering that have helped the economy grow. Pedagogy is the most important factor for producing the quality manpower for development in both countries.

FOR DISCUSSION

1. The Japanese method of pedagogy has a direct impact on the effectiveness of the performance of their students' economy. Do you agree?

2. Assess the effectiveness of problem-solving in mathematics education as strategy for preparing students to be workers in the economies of the United States and Japan.

Figure 4-1. The Panama Canal, Panama

Part 2

Theories of Learning and Principles of Teaching

Chapter 5

Philosophy: The Foundation of Pedagogy

From the beginning of time, man has been preoccupied with the phenomenon of learning. Man as a part of the universe must grow and change to maintain and perpetuate the progressive development of the human race as thinkers and workers in a viable physical environment. Learning is growing. How learning occurs has remained a mystery, but the most prevalent explanation has been through a phenomenon called *mind.*

Bases of the Teaching-learning Process

From the earliest civilizations knowledge has been transmitted to the young through teaching. Several explanations of how children learn from teaching are intuition, empirical research, theory, and philosophy. Philosophy has been the discipline that has postulated the most enduring and systematic explanation of how man learns and how to teach. It is philosophy that suggested that man learns through *mind.*

Intuition

The earliest explanation of how man learns was based on intuition. The theory is based on the idea that anyone can teach. As long as the person is in possession of the relevant knowledge, the individual can be a teacher.

This is based on the idea that a teacher is like a craftsman who can perform a skill intuitively, even without understanding the structure of the knowledge or the science and art of delivering that knowledge. In the view of the professional educator, teachers are artists who devote some philosophical thought to their craft, and then integrate this reflection into an art form (Henderson, 1996).

Empirical Research

Empirical research was first applied to learning when science touched philosophy and created a new discipline called psychology. Scientific methodologies such as observation and experimentation are applied to the teaching learning processes and have given rise to new methodologies in teaching such as behaviorism and cognitive theories.

Theory

Theory is a synthesis of philosophical thought and the findings of empirical research. As such it provides certain principles that may be used to explain learning. The importance of these theories is that they may be generalized from one situation to another and can be a paradigm for measuring the effectiveness of the teaching-learning process, methodology, assessment of truth, and discipline.

Philosophy

Philosophy is a systematic attempt to explain learning. Whereas intuition is a number of maxims, philosophy has a systematic method of interacting parts that must be consistent with each other. The parts are metaphysics, epistemology, and axiology, which are applied to the main concepts in education. Philosophy is the organization of thought based on logic and reason. It provides an explanation of reality, knowledge, and truth and is consistent with education as it studies concepts of mind and matter and their relationships to educational problems such as aims of education, curriculum, assessment, teaching, and discipline.

Philosophy: The Foundation of Learning Theories

An old maxim states that "a teacher has not been effective until he can teach Latin to John." There are two related concepts involved. John suggests that the person has a mind and Latin is the subject or what is called matter. Mind and matter are two concepts basic to an understanding of learning and to philosophy. An effective teacher must be the master of both knowledge and pedagogy.

Philosophy is a systematic attempt to explain learning. It consists of a specific organization of interacting parts that in philosophy are called metaphysics, epistemology, and axiology. *Metaphysics* examines ultimate reality and its relationship to mind, both of which are immaterial or conceptual. *Epistemology*, otherwise called the theory of knowledge, examines what is known, how it is known, and how to determine truth. *Axiology* examines the worthwhile values of ethics or right conduct and aesthetics, which evaluates what is beautiful. When applied to education these concepts have implications for the main problems in education such as educational aims, the curriculum organization, teaching methodologies, assessment strategies, and disciplinary practices.

Mental Discipline

The theory of mental discipline was first introduced by Plato and is based on the idea that the study of certain disciplines, particularly mathematics and the classics, enhanced mental functioning. In other words, school subjects taught with rigorous memorization discipline the mind. The theory has dominated schooling throughout the ages.

The theory of learning called mental discipline was derived from the philosophical positions of both Plato and Aristotle who in different ways both promoted the ideas about mind as the seat of learning. Mind was related to body, and the difference between the philosophers lies in how they conceptualized this relationship. Simply put, mental discipline means mind plus discipline, that is, mind plus knowledge. This section therefore analyzes metaphysics, which deals with mind, and epistemology or the theory of knowledge.

Metaphysics: Mind and Matter

The question concerning the nature of Mind and Matter is central to the theory of learning. It has its basis in a branch of philosophy called metaphysics and asks the question, "What is real?" The answer may lie in considering the work of a craftsman who made a table. A craftsman is concerned with the idea of a table before he can create a physical table. The table may be destroyed; it is mutable, and when it is destroyed the physical table no longer exists. According to Plato, however, the concept of the table lives on permanently.

Beginning with the creator of metaphysics, Plato, idealists claim the only reality is reducible to one element: idea. The table may be destroyed and thus the material table does not represent reality. Realists, beginning with Aristotle, who founded the philosophy, on the other hand, claim that both the idea and the material object are real. Such is the nature of the argument that began the immemorial controversy of the nature of ultimate reality, and its foundation in learning. Both Plato, founder of Idealism, and Aristotle, founder of Realism, thought that matter was separate and distinct from the principles of life, but came to different conclusions about the relationships.

Idealism and Metaphysics

Idealism is based on the idea that reality is reducible to one element. Idealists call that element spiritual, idea or concept. To Plato the idea of the table underlies the material table which is seen by the mind. The material table appears as real because man is limited by his senses, but the material is perishable. Ultimate reality is the idea which cannot perish. Man's nature is spiritual, and both material and idea are one, thus the Idealist are monists. Pragmatism is also a philosophy that embraces monism, but pragmatists believe that ultimate reality is material whereas idealists believe that reality is spiritual. Idealists are monists.

Realism and Metaphysics

Aristotle, founder of Realism and a pupil of Plato, disagreed with his mentor. He believed that both the idea of the table and the table itself have real existence. Idea is signified by thought; it is the soul that stands for perfection and struggles to maintain the selfhood of the individual and is therefore real. Each material entity has an immaterial form or idea. Both the material object and the idea together comprise reality. Thus reality is dualistic, and Realists are dualists. It is the material part of Realism that, when touched by science, gave rise to Pragmatism and the potential for the development of new knowledge

Mind and Human Nature

The basic purpose of understanding the metaphysical questions of human nature is man and the question of mind and its relationship to the body. The human mind is basic to understanding how knowledge is acquired. There are two views of human nature. Idealists, who are monists, claim that mind and matter are one and mental states are reducible to bodily ones. Because mind and body are one, mind is immaterial and comprises innate ideas, whereas body has transitory sense impressions. Mind knows when it is at one with the Infinite Mind or total body of truth. This is the theory of Idealism.

Realists, who are dualists on the other hand, believe that mind is an entity distinct from the body. Whereas the body is physical matter in nature, mind is immaterial, dynamic, and possesses free will, which makes it capable of thinking and making choices. Mind also has faculties that are abilities including locomotive, vegetative, appetitive, sensory, and reasoning. The sensory faculty responds to stimuli, and the reason analyzes and synthesizes them into new knowledge. The matter of how the body interacts with mind, however, is a mystery. There are several theories such interactionism, parallelism, occasionalism, and pre-existing harmony, however, dualists have not conclusively explained how mind interacts with the body. Realists are dualists, and both mind and matter are real.

Implications of Metaphysics for Educational Aims

Idealists and Realists have some common ideas. Both believe that mind and matter exist, and this leads to some common educational outcomes, such as the aim of education being mental discipline. The dissonant elements in Realism lead to Pragmatism, which regards only matter as ultimate reality.

The metaphysical positions about ultimate reality as idea and matter and the nature of human nature have educational consequences. Idealists, Scholasticists, and some Existentialists propose that ultimate reality is idea or spirit and that man is distinct from animals because of his spiritual nature. To these philosophers the spiritual nature that is present in every man suggests that the self is constant and thus remains the same throughout life, is eternal, and endures forever. Education is therefore permanent, touching eternity. The human mind is reasoning, thus reasoning must be the basis of education. The aim of education should be mental discipline or mind plus discipline. This discipline should be to liberate the human mind or to study abstractions at the highest level.

Philosophies such as Realism, Naturalism, and Experimentalism are based on the idea that both matter and mind are separate but real. Realists agree that both mind and matter are real whereas to pragmatists and Materialists only matter is real. Realists promote mental discipline as the aim of education and believe that knowledge is derived from experiences. Experiences, however, cannot account for knowledge. By use of reason humans may discover knowledge that exists independently of their perceptions; knowledge may be accessed by reasoning about experiences acquired through and sense impressions. Realists therefore arrive at the same conclusion as Idealists: the aim of education is therefore mental discipline, which literally means mind plus discipline or subject matter.

Epistemology: The Theory of Knowledge

The next branch of philosophy that addresses learning is epistemology. Epistemology asks three questions.

- What is known: the cognitive discipline?
- How does one know: the connection between the mind and knowledge?
- How is truth determined?

These three questions relate to curriculum, teaching methodology, and evaluation respectively.

Epistemology is the theory of knowledge based on two distinct philosophies. Plato conceived of learning through reasoning by accessing innate ideas in mind and in the intelligible world or the world of forms. Man has innate ideas that are in mind. Man learns through the interaction of reason, which is in mind, and the innate ideas, which are also in mind. Rationality is the theory proposed by Idealists who believe that ultimate reality is ideas in the *mind.*

Complementarily, Aristotle's school of Realism is derived from the theory of learning based on reasoning about experiences. Realists refute that there are innate ideas. Man knows through the senses when mind experiences the physical world and draws generalizations by reasoning about the experience. There was plurality in the world and, according to Koestenbaum (1968), for Plato the Good guided everything, but Aristotle viewed science was a nonhierarchical system by which to classify events.

Idealism and the Theory of Knowledge

Rationality is based on the monist principle that ultimate reality is idea and that the mind of man was born with innate ideas derived from the intelligible world. Plato also referred to the mind as the soul and held that to be the seat of all knowledge (Frost, 1962). Ideas were implanted in the soul but were clouded by the process of birth. The purpose of education is to help the individual to recall innate ideas. Knowledge was a priori or pre-existing, and learning is to recall of facts through reason.

How Man Knows

How man knows has been described by Plato in the *Allegory of the Lines*. There are two worlds: at the upper part of the line is the Intelligible World or World of Forms or Ideas, also called Reality and the World of Appearances or Phenomena.

In the World of Forms exists complete knowledge (epistome), which is a priori. Corresponding to the World of Forms is States of Mind or Intelligence (noesis), which acquires knowledge by reasoning or thinking (dianosia). Both are the upper part of the line. The worlds interact through consciousness; mind accesses the Intelligible World using reason. Truth occurs when what is seen in the physical world is consistent with facts in the Intelligible World, the Forms, or the Mind of the Infinite Good.

At the lower part of the line is the World of Appearances where there are visible material objects in the physical environment and corresponding states of mind. Mind can access the physical world using the senses. Consciousness links the world of appearances and mind using imagination (eikisia) and beliefs (pistis) in man's mind. At the upper end of the state of mind, there is the Intelligence.

In Plato's thinking, the World of Appearances is independent of the sensory world; it is ultimate reality that is nearer to the Good. Real knowledge is therefore a priori and is not derived from the world of the senses, but is absolute, constant, and immutable. Whatever is in the sensory world is always changing and is therefore inferior, whereas the ultimate knowledge has an intrinsic right to be called real. That is why Plato insisted that mathematics should be part of the curriculum of the Academy, the world's first university founded by himself. Mathematical concepts are immutable, and therefore provided training for the mind. Man should contemplate on the great constant concepts using reason; Plato's theory of knowledge was therefore rationalism. The Platonic theory was based on the Allegory of the Line (see Table 5-1).

Rationalist, from the Latin word *ratio* or *reason* persisted until the end of the Middle Ages and had a resurgence in the seventeenth century. The purpose of education was then to fill the mind with ideas and has been called Mental Discipline by the use of reason. Learning a course is by a process of rationality. Rote activities or memorization was the learning approach that dominated educational history from the time of the Greeks.

Kant, Hegel, and Descartes used methodologies, based on the theory of rationality and on deduction, the dialectic, intuition, authority, and linguistic analysis.

Table 5-1. Hierarchy of Knowing from Sensation of the Visible Object in the World of Appearances to Conceptual Knowledge in the Mind

REALITY	<consciousness>	MIND
The Intelligible World or World of Ideas or Forms (eido)		Intelligence (noesis)
Complete knowledge accessible by Reason	4.	Provisional knowledge (epistome)
Conceptual objects	3.	Thinking (dianoisa)
The World of Appearances or Phenomena		
(accessible to the senses)	^	Perception and
Visible objects >	2.	Imagination (eikasia)
	^	
Images	1.	Reflections and Beliefs (pistis)

What Man Knows

The mind of man is the same everywhere. It must learn ideas to liberate it. To attain mental discipline, curriculum must be based on essential subjects such as reading and writing, which are instrumental in nature. Study of the Liberal Arts liberates the mind to contemplate the highest level of abstraction. The curriculum is subject-centered, consisting of Liberal Arts subjects that reflect the cultural heritage. Human nature is unchanging; therefore, those subjects who were found to be good in the past are absolutely good.

How Truth is Determined

How truth is determined is the term used to describe when one knows that the information is correct. In Idealism, how man knows what is correct is based on the Coherence theory of Truth. Truth occurs when a real state of affairs is a mirror image of the world of forms, when factual knowledge is internally consistent (coheres). Idealists conceive of this knowledge as a priori, and constant, determined by comparing the answers of students with the established authority such as the textbook. Multiple choice questions ensure that answered are convergent.

Realism and the Theory of Knowledge

The opposite epistemology is the dualistic principle of ultimate ideas. Realism negates innate ideas, consigning the sources of ideas to interaction of the mind with the environment. Empiricism is derived from the Greek *empeirikos*, meaning experienced. Learning occurs when the mind receives stimuli from the environment via the senses. The mind analyzes the sensory stimuli thus received and organizes them within mind. Ideas develop in response to two types of sensory experiences: sensation and reflection. Sensation receives the stimuli and refection refines them into ideas. Learning is empirical, and knowledge is a posteriori or based on experience.

How Man Knows

Aristotle rejected Plato's theory of Forms. For Aristotle matter and Form are merely abstractions, that is, in a marble statue of Socrates, the marble is the matter and the shape of Socrates is the Form) they are both found in the object. Man knows when the mind apprehends the form of the objects, remembers it, and classifies it. To Aristotle knowledge is a posteriori, occurring after the experience and is dependent on it. This is different from Plato who said knowledge is a priori, that is, stemming from what had been learned prior to the birth of the soul.

Aristotle, a pupil of Plato, was the founder of the school of Philosophy called Realism. He conceptualized the Mind as having faculties, vegetative (responsible for growth), appetitive (sustaining life), locomotive (for motion), sensory (for connecting the individual with the environment), and rational (for analysis, discrimination, and synthesis of sensory experiences), thereby producing new knowledge. Knowledge is a posteriori, occurring by experience. Table 5-2 illustrates this scheme.

Based on Aristotle's theories, learning occurs when the mind receives stimuli from the environment via sensations and reflects on them. The rational faculty analyzes and structures the sensory.

Table 5-2. How the Mind Learns

Faculties of Mind	«———»	External Reality
Rational		
^		
^		
Sensory	«———»	*Object
Appetite		
Locomotor		
Vegetative		

Learning is based on empiricism. Stimuli from the environment enter the Mind. The mind analyzes and synthesizes them into knowledge and then stores them in the unconscious, which organizes them into knowledge. Ideas are developed through sensory experiences and reflections; knowledge is empirical and a posteriori.

What Man Knows

Knowledge is derived from sense experiences. Reality, however, is dual: both ultimate forms and those that are a posteriori may be apprehended by the senses. Sense experiences alone, however, cannot constitute knowledge. By using reason, man is able to discover the object or objects that exist independently of his perception. He may discover truth when the mind of man corresponds with extra-mental reality. Education must enable the individual to fulfill his potential through reasoning and also to transmit culture.

How Truth Is Determined

Dualists practice the Correspondence Theory of Truth, which occurs when a mental event corresponds with a factual state of affairs.

Educational Implications of Epistemology

Epistemology is concerned with curriculum methodologies of teaching and methods of assessment. Epistemology of Idealism and Realism produced a subject-centered curriculum. Methodologies were reasoning, authority, and activities that led to memorization. Assessment was by traditional testing.

Transitional Philosophical Learning Theories

The theory of Mental Discipline derived from the Philosophies of Idealism and Realism that dominated the learning world for centuries. They were to be refined during the seventeenth century by giants in the field, especially by Rene Descartes and John Locke respectively.

Idealism gave birth to Rationalism and Realism to Empiricism. Idealism was born in the fourth century B.C.E. and Realism in the third century B.C.E. During the Middle Ages, Platonic Idealism was the dominant philosophy because it was consistent with Christianity and the philosophy of Scholasticism. Aristotle's philosophy was quiescent during that time but was rediscovered during the Islamic expansion into Europe. It was then that its influence resurged and breathed new life into the prevailing ideas during the Renaissance. It raised questions about the nature of the universe and the scientific method of examining matter.

Scientific development reached full flowering during the European Enlightenment of the seventeenth and eighteenth centuries. It was during this time that Idealism and Realism were transformed to Rationalism and Empiricism. Idealism and Realism were refined by Descartes and Locke respectively. Both Descartes and Locke reflected on the works of Plato and Aristotle and were touched by the European Enlightenment and by science. They came to different conclusions, however, about the nature of reality and knowledge.

Rationalism

Descartes (1596–1650) accepted the Platonic principle of innate ideas and the doctrine that knowledge may be gained from basic principles in mind. He abandoned appeals to the authority, such as the Scriptures, and focused exclusively on reason. Descartes proposed that knowledge may be developed by deductive reasoning from a few basic ideas. The use of deductive reasoning or questioning was the means of accessing these ideas.

The purpose of education was to fill the mind with ideas, called Mental Discipline, by using reason. A mathematician and inventor of analytic geometry that bears his name, Descartes believed that mathematics was a subject that could prove his point. This reliance by Descartes on rational thought as the basis of knowledge gave the name Rationalism to a theory of knowledge. Thus Idealism stressed the study of ideal forms of knowledge such as mathematics. Mathematics is a good example of how knowledge is derived from a few basic principles through deductive reasoning.

Empiricism

By contrast Locke (1632–1704) rejected the principle of innate ideas. He relied on the Aristotlean principle of learning via the senses and was touched by the burgeoning scientific revolution of the Enlightenment.

Empiricism was the name given to his theory of knowledge. Locke proposed that at birth mind was a *tabula rasa* with a chemistry that allows it to compare and contrast and that associates impressions. Mind accumulates and organizes ideas within an individual's sensory experiences. Thus whereas Realism stresses that knowledge is gained via the senses, and mind structures and organizes sensory experiences, Empiricism stresses that ideas are developed via two types of sensory experiences, sensations and reflection (Gredler, 1986).

Locke lived during the age of Newton when science was developing rapidly and many thinkers contend that he was concerned with developmental psychology (Hamlyn, 1987). Indeed Locke was a doctor of medicine and during this period new inventions and ideas abounded. According to Frost (1962), Locke said that "as Creator of the world and man, has established certain laws which man may discover through studying nature or revelation" (p. 30).

Locke was clearly a transitional figure. He was at once influenced by the observational science of Copernicus and also was constrained by ideas of the dualism between mind and matter. His Empiricism as a theory of learning was a foundation on which later psychology was built. Locke's work was *Concerning Human Understanding* (see Burnet, 1984).

Unfoldment of Mental Capabilities

Jean Jacques Rousseau (1712–1778) wrote *Emile*, the education of a boy in a natural setting, and subsequently proposed Naturalism as the theory of learning. Rousseau further developed Locke's ideas. Rousseau suggested that man's nature was basically good; it was society that corrupted it. Education should take place by living close to nature, the highest good. Learning was *unfoldment of the natural tendencies inherent in the individual* by experiences derived from interactions with the environment.

This theory was furthered by the Swiss educational reformer, Johann Heinrich Pestalozzi (1746–1827) in his book, *Leonard and Gertrude*. His protégé, the German Friedrich Froebel (1782–1852), was the founder of the kindergarten and the concept of learning through the natural tendencies such as play, as he expressed in his book, *Education of Man*. The theory promoted a developmental approach. Learning would occur by close interactions with the environment. Maturation and the needs and interests of children influenced behaviors and learning.

Apperception

Johann Friedrich Herbart (1776–1841) was both a philosopher and psychologist. In his books, *Science of Education* and *Outlines of Educational Doctrines*, he expressed his philosophical theory that the universe consists of great number of unchangeable principles or substances that he called *reals* (Frost, 1962). As one rearranges reals, one creates experiences. Interaction among reals produces reorganization. Man as part of the universe is governed by fixed laws, man can be explained by these laws, and learning occurs as one experiences these laws. By recognizing this phenomenon, Herbart was laying the foundation for the application of scientific method to education.

Herbart was the founder of Associationist psychology. His theory of Apperception stated that learning was accomplished by associations. When simple sense impressions from the external world are associated with other perceptions, it creates a mass of perceptions in the mind. He described the mind as a neutral and passive storehouse for ideas, or mental states, and thus as the totality of mental states or an Apperceptive Mass.

Mental states, to Herbart, are sense impressions, images, or copies of previous sense impressions, and emotional states. They are the only source of mental activity and are called presentations. The mind is empty or passive until the first presentation, and new concepts are learned by association of the new and existing ideas in the Apperceptive Mass. Perceptions were also stored in the unconscious and recalled with the occurrence of a newly associated experience. The mind focuses on only one idea at a time.

Herbart suggested that learning begins with accessing existing mental states via sense activity, followed by presentation, association, and conceptual understanding or generalization, and ending with application.

Herbart developed an instructional model that was consistent with these principles and based on his Associationist Theory, which became famous in Europe and North America. The empirical tradition became important in education practice and gave birth to psychology where it was evident in the works of John Dewey, Maria Montessori, and Jean Piaget.

Experimentalism

Experimentalism is the group of philosophies that emphasize empirical verification of truth. These include the philosophies of Pragmatism proposed by William James (1842–1910) and Instrumentalism proposed by John Dewey (1859–1952). Experimentalism is also called Progressivism.

William James built on the works of J. S. Mills (1908–1873) and C. S. Pierce (1839–1914). James's theory was that a "block universe in which everything was governed by the laws of the universe, and that the test of every theory was its practical consequences" (Frost, 1962, p. 50). This was the pragmatic test. The real world was the world of human experiences, and knowledge can be explained by terms of the relation of one part to another (Hamlyn, 1987).

John Dewey (1859–1952) was a phsychologist and philosopher as well as a leader of the school of Pragmatism. His proposal that the world is ever changing, growing, and developing strongly reflected the influence of evolution. Dewey focused on experience, which was always becoming and being enriched. His theory, Instrumentalism, proposes that education is like a tool, an instrument.

Pragmatism

As a philosophy, Pragmatism is derived from Realism, which shows a dualism between mind and matter. Pragmatism is monist, representing the parts that matter only: man is a part of the universe. Man is part of the world of perception and the world perceived. All he knows is based on beyond his experience. It is antimetaphysical, as metaphysics does not lend itself to experimental verification. Mind is not an entity; it is instead a process by which man adjusts to the environment. Ideas are plans of action, and spring from the functioning of the brain. By conceiving of mind as process, pragmatism showed the influence of evolution.

As an epistemological theory, Pragmatism believes that the source of knowledge, what man knows, comes from experience. Because the senses may be misleading, sensations must be subject to verification, and the test of a proposition must be social verification or that which the community determines to be worthwhile. Human experience begins with a problem. Man has to think to solve the problem; therefore ideas are plans of action, and thinking is a process by which man solves problems. Knowledge is the outcome of the problem-solving process. Knowledge is posteriori.

Because the world is ever changing, knowledge is also changing; there is no absolute. The method of transmitting knowledge is via the scientific and the problem-solving methods. Truth is determined by the pragmatic theory of truth that is by verification, that is, whatever works. What is true is known by testing it against reality, which is an act. Evaluation therefore reflects the modern method of authentic assessment and performance evaluation.

The educational implications of Pragmatism are that the aim of education is to help man to adjust to the environment. It is problem-centered. Man must learn how to think, not what to think. The curriculum must exist in a social context. The subject matter is a tool for solving problems and also an outcome of the problem-solving process. It should not be based on experiences of the past, which should only be used as a tool or an instrument. The curriculum should lead to the reconstruction of the individual and the society.

The philosophy of Pragmatism has also been described as a progressive educational theory. The methodology of teaching has been called nonauthoritarian, psychological, and problem-centered. The curriculum has been described as experiential and student-centered and the teacher as nonauthoritarian or a facilitator.

Conclusion

From the earliest Greek civilization, philosophy was the main source for explaining natural phenomena and was the basis of learning theory. The philosopher Plato (427–327 B.C.E.) was the founder of Idealism, which describes basic reality as innate ideas that are inborn. His pupil Aristotle (384–322 B.C.E.), the founder of Realism, insisted that ultimate reality was both mind and matter. These ideas gave rise to the theory of mental discipline. Idealism and Realism were later refined by Descartes and Locke respectively and gave rise to the epistemological methods of Rationalism and Empiricism. Rationalism was based on the concept that knowledge could be derived from a few basic ideas, and Empiricism contended that learning was based on sensory experiences that were analyzed by reason.

Empiricism was succeeded by transitional philosophies such as unfoldment, apperception, and associationism, which are based on the principle of learning through experiences. Pragmatism was developed in the twentieth century by James and Dewey, who emphasized that learning began with problems. The child-centered approach to pedagogy was the basis of this philosophy.

ASSIGNMENT

Read the chapter and the references and then answer the following questions.

1. Describe how metaphysics translates into the aims of education.

2. Discuss the implications of mental discipline as an aim of education.

3. Describe how man learns according to (a) Idealism and (b) Realism.

4. What is truth and how is it determined? What are the implications for the classroom?

5. Discuss Rationalism and Empiricism as pedagogy.

Figure 5-1. Aztec Calendar

Figure 5-2. Aztec Temple at Teotihaucan, Mexico

Chapter 6

Psychology and the Development of Learning Theories

From 400 B.C.E. to the mid-nineteenth century, philosophy served as the primary source of information about the human mind and learning (Gredler, 1986). For centuries, man relied on the authority of Aristotle, Ptolemy in astronomy, and Pliny and Elder in the natural sciences for scientific knowledge. During the thirteenth century, there were stirrings of the forthcoming divorce of science from theology, when theories of the universe were challenged, and these new ideas emerged into the scientific evolution of the eighteenth century. It was then that the principles of science were applied to philosophy and gave rise to psychology.

The Influence of Science Transforms Philosophy to Psychology

Science was to change the course of events that influenced learning theories. It was to transform philosophy in two fundamental ways. First, the observational and experimental science of Copernicus, Bacon, and

Galileo created the concept that mind has a structure on which experiments could be performed using experimental research methods and from which learning theories could be developed. Second, the evolutionary science of Charles Darwin in his theory of evolution replaced the idea of dualism of mind and body with the concept that mind is a process. This gave birth to cognitive science. The progress was in developmental stages.

The greatest advance came in 1543 when Copernicus (1454–1504) disproved the theory of the geocentric view of the universe that had been proposed by Ptolemy in support of religious doctrine based on observation. Copernicus proposed a daring hypothesis. The sun, he concluded, was the center of the solar system, and the earth was but one of a number of planets that revolved around it at varying distances, rotating on their axes. Copernicus based his heliocentric theory on scientific method of observation. He was a mathematician and an astronomer and did most of his observation indoors from his study. Copernicus' work was published in a book, *De Revolutionbus Orbium Celestium* (Concerning the Revolution of the Heavenly Bodies). It was placed on the Index of Prohibited Books by the Church, where it remained for 150 years.

The next advance was with the work of Francis Bacon (1561–1626), who introduced the concept of inductive reasoning. Bacon, in his *Novum Organum* and *Advancement of Learning,* set out to determine the correct methodology for acquiring knowledge (Hamlyn, 1987). According to Frost (1962), Bacon laid the foundation for modern science. Having relegated religion to a realm all its own, he set about developing a method of thinking that he believed would give true knowledge about the universe. This method is called *induction.* By the study of likenesses and differences among things, man could discover the laws, causes, or *forms* of objects in the universe. In this way one could understand the universe (Frost, 1962). The universe, according to Bacon, exists and operates in accordance with fixed laws, and levers serve as keys to unlock the universe. Bacon felt that "there was method for eliminating all but one of the finite number of possible forms for a given phenomenon, and that the method could lead to certainty" (Hamlyn, 1987).

The next great advance came with the work of Galileo Galilei (1564–1642), who introduced experimental science. Before he was twenty Galileo discovered the law of the pendulum, which stated that the successive swings of the pendulum were of the same duration. He investigated the law of motion and solved the problem of falling bodies in terms of uniform acceleration as distinguished from uniform velocity. Galileo became a convert to the Copernican ideas but did not champion it until he had made more discoveries with the telescope. He constructed a telescope and became the first scientist to apply the telescope to scientific studies. He discovered that the moon had no light but owed its light to the sun and that the moon had furrows. He discovered the four satellites of Jupiter, the spots on the sun (inferring that the sun moved on its axis and therefore rotated around the earth). In 1632 he published his work under the title *Dialogue Concerning the Two Chief Systems of the World*, which presented proof of Copernican theory. For this Galileo was forced to recant. Historians regard Galileo as the founder of experimental science. His investigations of nature discredited the method of reliance on authority for knowledge. A notable era of scientific development followed Galileo's inventions, including Isaac Newton and the law of gravitation, William Harvey and the circulation of the blood, and Zacharias Jensen, a Dutch spectacle-maker, and the microscope.

Charles Darwin (1809–1882), a British scientist, proposed the concept of evolution in his work, *On the Origin of Species by Means of Natural Selection* in 1859. The thesis described in the title was that organisms had evolved by minute mutations out of more elementary forms by natural selection. The impact was that it focused public attention on the processes of change. It defined reality as given to change rather than static. The accepted view of human nature changed, and the idea of a dualistic view between mind and matter was overturned. It was replaced by the idea that mind was not an entity, but rather part of the evolutionary process.

Influence of Observation and Experimental Science

The foregoing discussion provides background for scientific development and its potential for influencing education. This section describes the synthesis of philosophy and science. Scientific method was the immediate outcome of the merging of scientific methods involving observation and inductive thinking proposed by Bacon. When experimental science, the idea of performing experiments and observation, proposed by Galileo touched scientific empiricism, it gave rise to ideas about experimentation, stimulus-response theories, and then behaviorism. By contrast, evolutionary science proposed by Darwin directly influenced ideas of learning as a process and then neuroscience.

Scientific Empiricism

Locke's Empiricism, which was elemental science, was influenced by the scientific observation of Copernicus; the observation, experimentation and inductive reasoning of Bacon; and the experimentation of Galileo. The interaction of these influences gave birth to Scientific Empiricism. It inspired Herman Von Helmholtz, a medical doctor, scientist and philosopher who refuted Descartes' argument of innate ideas and corroborated Locke's idea that knowledge was derived through sense experiences. He concluded that if ideas are the product of mans' experiences, then they are subject to observation and analysis. Thus Von Helmholtz coined the term *scientific empiricism*, a mixture of empiricism and observational and experimental science.

Based on these assumptions, Von Helmholtz proposed that accumu-lated experiences could be examined by controlled experiments from which new knowledge could be attained. Von Helmholtz demonstrated this theory using experiments on the senses. He invented the ophthalmoscope, a medical instrument for observing the functions of the eye.

Structuralism

Following research on the sense organs, which are living tissue, the next step was to perform research on the mind itself. According to Platonic thought, the mind is the seat of ideas. Aristotle defined mind as having faculties or abilities, including reason and the sensory faculties (taste, hearing, sight, smell, and touch). The mind is able to transform sense experiences into ideas and store them in the form of ideas. Thus Wilhelm Wundt perceived mind as an immaterial entity on which experiments could be performed by examining its basic characteristics.

Wundt employed a typical method used in chemistry that identified molecules and atoms as the basic elements of the matter. He translated this method by defining the manifestation of mind operationally and measuring the indicators of mind, which he said had structure or characteristics. Thus he conducted research on characteristics of mind such as sensation, auditory perception, and attention. By measuring the structure of mind, Wundt introduced a methodical approach to psychology called Structuralism.

In 1874, Wundt published his research on sensory functions in and named his methods "Physiological Psychology" based on the functions of the body that inspired his research. Based on this methodology, Wundt, a philosopher and scientist, became the father of experimental research commonly used in psychology today.

Structuralism became an established a branch of psychology. It was furthered by other researchers, chief of whom was German psychologist Hermann Ebbinghaus, who invented the nonsense syllable that used sensory reactions to measure learning. Structuralism sought to develop a pure science based on the human mind. The problem was that it relied on the respondents to give self-reports of their thought processes, which at best led to unreliable results. Further, psychologists could not agree on how to measure manifestations of the senses such as imagery and feeling. Moreover Structuralism did not accept the idea of evolution, although the majority of researchers found it acceptable. Thus in 1927, Structuralism died a natural death with its last champion, Edward Titchenor.

The Influence of Evolutionary Science

Evolutionary Science gave new meaning to the concept of mind. It conceived of mind as a process rather than as an entity with a structure on which to perform experiments. It was a developmental process with inputs, process, and outputs as part of the same system. It represented growth, which begins with solving problems in the environment and creating new knowledge. As in evolution, Evolutionary Science regards the mind as struggling to survive in the environment and constantly adjusting: a dynamic process.

Functionalism

John Dewey (1859–1952) was the creator of applied psychology, which he called Functionalism. He was also a philosopher and in both these disciplines he demonstrated the influence of Evolutionary Science. Dewey criticized Wundt's method of structural analysis, which he called the psychology of classification. He compared it to the practice of biologists in classifying organisms into taxonomies and said that if psychology were to follow the method of biology, then it should move beyond the method of taxonomies to the study of processes.

Process then was central to Dewey's analysis. He suggested that Wundt's division of behavior into "stimulus and response" or "stimulus and movement" was artificial and not based on reality. In reality, stimulus-and-response syndrome was part of a process or behavior that Dewey called *Functionalism*. His teachings led to the development of an experiential curriculum in public school (Bell-Gredler, 1986).

The Growth of Learning Theories: From Stimulus-Response to Neuroscience

Learning theories developed during the early twentieth century were different from philosophical theories that were based on intuition and logical analysis. Psychological theories were all based on scientific principles, that is, observations of behavior and experimentation. The researchers applied hypothesis and scientific principles to real-world situations followed by generalizations about learning in human beings.

Stimulus-Response–Associationist Theories

During the twentieth century psychologists moved away from Structuralism and Functionalism. They synthesized Herbart's version of Associationism and Wundt's experimental Structuralism to create a more physiologically based theory that was both experimental and based on associationism. These were Stimulus-Response (S-R)–Associationist Theories. Among the theories in this group are Thorndike's Connectionism or Instrumental Conditioning, Pavlov's Classical Conditioning, Guthries's Substitution or Contiguity Theory, Hull's Reinforcement Theory, and Skinner's Operant Conditioning. Eventually, a synthesis of Classical Conditioning, Connectionism, and Functionalism resulted in Behaviorism.

Classical Conditioning

Ivan Pavlov (1849–1936) took the lead in developing what has been called *Classical Conditioning.* In 1904 Pavlov received a Nobel Prize for his work. The experiment consisted of giving meat to a dog and simultaneously ringing a bell. The dog would respond by salivating. This was an unconditioned response. The next event was to ring the bell, without giving any meat. The dog would salivate. This was conditioned response, or a learned response; the dog associated salivation with the sound of the bell.

If stimuli other than a bell were used, the dog would salivate, that is, the response would be generalized. The experiment established principles of learning including

- *Contiguity*: The stimuli must be presented close together;
- *Repetition*: Stimuli must be paired several times before conditioning will occur, and;
- *Motivation*: The subject must be hungry.

Thus in the classroom stimuli must be presented together, several times, to students who are motivated or interested in learning.

Connectionism or Instrumental Conditioning

Thorndike (1896) followed Pavlov's lead in performing experiments on animals, cats, dogs, and chickens. His dissertation was a landmark event that demonstrated how the conditioned response may be instrumental in satisfying some need of the individual. It had as its basis the experimental works of Ebbinghaus combined the psychomotor skills acquired by learning telegraphy (Boring, 1950). He chose animals as his subjects so as to study the "Mind of the Animal," according to Thorndike, to discover the development of mental states, particularly the development of intelligence.

Thorndike was clearly applying experimental research methods of Structuralism to the concept of Process, which was the basis of Functionalism. His method was to measure a sequence of external events, behaviors, and consequences in order to make inferences about mental state. It was an experiment in Stimulus and Response (S-R) or Associationist theory. Thorndike's experiment produced the formulation of a learning theory. His experiment was based on a cat placed in a puzzle box with a piece of fish placed outside the box. The experiment was for the animal to escape so as to get the fish outside. From this he developed three laws: of exercise, effect, and intensity.

The *law of exercise* embodied the principle of repetition and readiness for association applied to the situation and response. A response becomes associated with a stimulus.

The *law of effect* was related to the satisfactory or unsatisfactory response. When a satisfactory response closely follows a stimulus, it will likely recur. The law of effect indicated that the cat's hunger was its mental state (Lovell, 1973). The law of effect defined the state of affairs that will establish whether the connection will be made. When a connection is made between the stimulus and the response, and satisfaction follows, the strength of the connection is increased; when it is followed by unsatisfactory results, the connection is weakened. Thorndike therefore showed the association between the external stimulus and the corresponding mental state. Thorndike's theory is also called *instrumental conditioning* because the selection of a particular response is instrumental in obtaining the reward.

The *law of intensity* states that the greater the satisfaction derived from the experience, the stronger will be the bond between the stimulus and the response.

Substitution or Contiguity Theory

Guthrie rejected Thorndike's law of effect, which emphasized the bonds being strengthened when they were followed by a satisfying state of affairs. His theory is an associationist theory based on the connection between the stimulus and response. He theorized that all responses are associated with the stimuli, and that when the stimulus is repeated the response will be repeated with full association being established on a single pairing of the events. Repetition has no effect. Forgetting is the result of interference by subsequent stimuli. Reward does not enter into the learning equation. Contiguity in pairing the stimuli, however, is important and motivation is necessary to initiate activity but not otherwise. According to Guthrie, the teacher should not concern himself with motivation, or reward. He should remove all distracting influences.

Reinforcement Theory

Reinforcement Theory by Clark Hull (1943) furthered Thorndike's theory. He experimented with a rat placed in a maze. His approach applied the principles of stimulus-response but added the new dimension of reinforcement or reward and its power to reduce the animal's needs. Learning would only occur when the individual's need was reduced or satisfied. Further, if the response was strong, it would recur at a later time; each time the stimulus occurred, the number of errors would be reduced. Each mistake would be penalized by denial of food, and reward strengthened the response. After several tries, the rat would succeed. Reinforcement made a contribution to learning theory.

Hull's theory is systematic. He emphasized initial intervening and output variables, or a sequence of events that results in the formation of habits. The stimulus called a *drive* led to an impulse in the individual's response, which reduced tension. The occurrence resulted in the reorganization of the nervous system, that is, a habit. An organism thus builds habits by responding to stimuli in the environment.

In sum, Associationist theory is built on the idea that learning takes place when the bonds between stimulus and response are strengthened. Strengthening occurs when the stimulus and response and reinforcement occur simultaneously. Responses that are followed by satisfaction or reduction in the mental states of the subject are repeated, and those that do not produce satisfactory results are not repeated.

The Rise of Behaviorism

The methods of Classical Conditioning (a link between stimulus and response) and Connectionism (a behavior, an indicator of a mental state) were fused into a new method called Behaviorism. The idea was that a behavior could be conditioned.

Classical Behaviorism

Launched by Watson (1879–1958) in 1913, it became important during the 1920s. Watson's aim was to study the behavior rather than the mental state. He started with the premise that all organisms adjust to the environment, therefore behaviors rather than mental states should be studied. According to Watson, because certain responses follow certain stimuli, psychologists should be able to predict the response from the stimulus and the converse. There could therefore be experiments on behaviors.

Watson defined mental states as behavioral responses. For example, he defined thinking as subvocal speech and feeling as glandular reaction. He believed that human behaviors and personality were the result of conditioning, which would allow experiments on behaviors. Watson was able to condition an eleven-month-old boy Albert to be afraid of furry objects and then to uncondition him. The essence was that Watson conditioned an emotional state as a reaction to a stimulus.

Operant Conditioning

Following in the tradition of Watson was Burrhus Frederic Skinner (1904–1990). According to Skinner, the goal of any science is to discover the relationships among natural events in the environment; therefore, the science of behavior must discover the functional relationships among physical conditions or events in the environment and observable behavior. Like Watson, he believed that psychology can become a science through the study of both the dependent and independent variables of behavior. Skinner defined learning as a change in behavior. His form of Behaviorism was known as *Operant Conditioning*. The stimuli would be reinforced. He derived his idea of reinforcing from Edward Thorndike's law of effect. The effect of the subject's action would cause behavior. Skinner in effect synthesized the methodologies used by Thorndike and Pavlov (Bell-Gredler, 1986).

Skinner was concerned only with observable behavior, an operant (that is, a set of acts that constitute an organism's responsible pattern or behavior pattern), for example lifting an arm or a monkey pushing a lever. His theory was that learning is a change in the probability that a particular response will or will not recur. An operant is available behavior that is conditioned by reinforcing it and that is likely to recur in the future. A subject will repeat the reinforced or rewarded behavior that Skinner called *reinforced consequences*. Thus operant conditioning is the process that makes a response probable through the use of positive or negative reinforcers. A positive reinforcer, such as food, strengthens the behavior. A negative reinforcer, such as a loud bell or an electric shock, weakens the behavior.

Skinner performed experiments in a darkened laboratory because he was concerned with a simple environment with few stimuli. He used devices such as levers, keys, and discs. Reinforcers were food and water, and stimuli were light, loudspeakers, and an electric shock. He recorded the responses on automated devices. Skinner used a box, now famously called a Skinner Box, in which an animal was presented with a lever (a stimulus). Whenever the animal pressed the lever (an operant) food was released. Food was a positive reinforcer, and strengthened the behavior so that the behavior (pressing the lever) was likely to recur. Skinner's goal was to control the environment so that he could predict behavior. He was in fact conditioning the animals in the experiment. The method became famous as behavior modification, a method of controlling subjects. Criticisms of his experiment included that human behavior was more complex than that of animals (Smith & Smith, 1994).

Cognitive Theories

The ability to practice experimental research on learning by behaviorists was impressive, albeit that these experiments were performed on less complex creatures such as animals that were already less complex than human beings. Critics identified two shortcomings, however: one dealt with the environment and the other with the simplicity of the stimuli. First, the environment has to be controlled such that it has only one stimulus.

This condition does not exist in a realistic situation such as a classroom. Second, the emphasis was on external behaviors rather than cognitive processes. For instance, the number of times that a rat in a maze ran up a blind alley was regarded as errors by the researchers. Edward Tolman (1948) regarded these not as errors but as exploration that made the rat aware of its environment and at a later date helped guide the rat to choose the right path, thereby building a cognitive map.

Cognitive theorists criticized behaviorists because they concentrated on external forces and did not take into account what happens inside the mind of the individual, as cognitive theorists did.

Purposive Behaviorism or Sign-Gestalt-Expectation Theory

Edward Tolman, author of this theory, wrote his thesis in his work, *Collected Papers in Psychology* (1948). He rejected the stimulus-response theory of the associationists. He conducted a number of experiments with three groups of rats running a maze on ten successive days. At the end of the run, one group received food and another group received none. The group that received no food on each successive run received a large quantity of food on the eleventh day. The error rate of this latter group dropped thereafter such that it was no different from the group that received food at the end of each run. The inference was that reward for food did not affect performance or latent learning. Tolman also did experiments to prove cognitive learning and maintained that rats learned cognitive maps of the environment. Tolman's conclusion was that purpose is important in learning; the route was what is learned. Learning is recognizing relationships or signs that allow the learner to reach a goal. Rewards are not important; satisfaction comes when the goal is attained and is intrinsic.

Field or Topological Theory

Kurt Lewin was the founder of what has been called *Field Theory*. He explained his theory in his work, *Principles of Topological Psychology* (1936). The basis of this theory is that the behavior of an individual operates within a field of forces such as demands, attitudes, sanctions, and attractions. These constitute the field of forces within the environment. Within this environment individuals have a life space that consists of people and objects encountered, private thoughts and feelings, and internal tensions and needs. The forces acting on the individual are called *psychological forces*. Lewin used rats in a maze to perform his experiments. He was primarily interested in motivation. Internal to the individual is a cognitive map that comprises the internal needs and past experiences of the individual or what was previously called *mind*. All behavior is purposive or goal-directed. The interplay of these forces, those in the environment and those internal to the individual, resulted in differentiation of the individual's life space, or learning. Repetition resulted in more differentiated knowledge. This theory depends on intrinsic motivation, which is the tension within the individual. Motivation is rewarded when the individual attains a goal and is satisfied. The emphasis on internal tensions, needs, and cognitive structure make the theory cognitive in nature.

Gestalts

Gestalt psychology was the brainchild of Max Wertheimer, Kurt Koffka, and Wolfgang Kohler. They rejected the idea that learning is a set of connections between stimuli and responses and articulated that learning was a set of mental processes and the key to understanding. Gestalts reacted against the tendency of Behaviorists to analyze experience into parts; they saw the parts as part of a single experience. According to Koffka, the basis for learning is that subjects react to unitary, meaningful wholes (Koffka, 1935).

Wertheimer proposed the laws of perceptual reorganization that are basic to Gestalt theory of learning. Gestalt began accidentally when Wertheimer had a hunch and bought a stroboscope, a precursor of the movie camera. The principle was that the exposure of two lines in quick succession, if carefully placed, are perceived as movement. In other words, the perception of the whole is different from that of its parts. This was the basis of Gestalten theory of perception as the basis of learning. What the mind does in the learning process is to reorganize the perceptual field. The laws of perception apply to learning. Wertheimer postulated the laws as follows:

a. The sensory field is organized into ground and picture The ground is indefinite, formless, and without significance, whereas the picture is the part on which the individual focuses attention.
b. The visual field is organized by the following process of proximity, similarity, open direction, and closure.
   Proximity is the nearness of the objects to each other,
   Similarity is shared features,
   Open direction is the tendency of objects to form a pattern, and
   Closure is the tendency of the perceiver to see whole objects rather than parts. Objects tend to be closed and in a complete pattern. Meaning occurs only when they are complete.

Wertheimer compiled his theories in his book, *Productive Thinking*, which was published posthumously. In this book he applied his theory to the solution of problems of parallel lines.

The concept of *Insight* or sudden understanding of the respondent to a given situation was the contribution of Wolfgang Kohler. He conducted experiments with anthropoid apes while on assignment by the Prussian Army in the Canary Islands. The experiment consisted of placing an animal in a cage. Outside the cage was a piece of stick and there was food nearby. The requirement was for the animal to take the two pieces of stick and get the food. Kohler found that on a first attempt the animal was able to perform the experiment. He referred to Thorndike's experiment and noted that if the animal had been able to view the whole situation at once, he would be able to solve the problem. Learning, Kohler theorized, was about problem-solving. The individual who was able to view the whole situation was able to solve the problem more efficiently. Learning consisted of observation of relations, reasoning, and generalization, or what the Gestalts call *insight*.

Gestalts regard trial and error as a necessary part of the process in the reorganization of the psychological field. Each time the action is done the field is reorganized, and the individual is able to view the entire field in another way; that is, the animal is setting of a series of hypotheses or cognitive maps. An important part of insight is that it may be reproduced in a new situation and transferred from one situation to another. Thus learning is generalizable.

Kohler also developed another concept known as *The Law of Transformation.* This time he did experiments on hens. He presented the hens with grain on paper of two shades of gray, but they were only permitted to eat from the darker shade. After conditioning, the hens aware again presented with grain from paper of two shades of gray: the original darker shade and a new darker shade. The results were that the hens ate not from the dark shape from which they had previously eaten but from the new darker shade of gray. The hens had learned that gray is relative to brightness. Absolute gray was not important. This law has been referred to as *transformation.*

In the teaching-learning process, knowledge has to be organized, stimuli be presented close to each other, be based on patterns, and be problem-based for students to gain insight or understanding.

Developmental Learning

Jean Piaget (1896–1980) followed in the tradition of empirical theorists, especially Rousseau, with his stages of human development. Like all cognitivists, Piaget was interested in the central organizing processes that shape human intellectual development (Smith & Smith, 1994). Piaget believed that cognitive functions help the individual to adapt to the environment. These functions are assimilation and accommodation. An individual needs a schema by which to interact with the environment. Piaget proposed that mental development proceeded in stages from the preoperation, concrete operational, and formal operations. The preoperational stage included manipulation of the environment, concrete operations are visualization and classification, and formal operations are abstract and logical operations. The stages phenomenon foreshadowed information-processing theory.

In summary, Stimulus-Response-Associationist Theorists are concerned with how the environment stimulates the individual and how the individual responds to the stimulus. Their successors, the Behaviorists, are concerned with how the individual reacts to stimuli.

By contrast, Cognitive theorists such as the Gestalts are empiricists who emphasize the oneness of the perceiver and the thing perceived, the object and the meaning. They believe that the organism and the environment constitute a whole; their interaction is mutual and brings about experience.

Cognitive Science and Neuroscience

During the latter half of the twentieth century, cognitive theorists have been influenced strongly by computer technology. Whereas experimental science and evolutionary science were predominant is the previous century, the twentieth century saw the beginning of a dramatic revolution in science. The nineteenth century was the age of physical science; the new century is the age of biological science. Physical science has given rise to computer technology and information processing. In biology and medicine have prompted medical research on the brain. Today the synthesis of these two phenomena has influenced how educators see learning. The new concept of learning has been called *cognitive science.*

Cognitive science is an emerging field that includes research from neuroscience, neuropsychology, developmental psychology, educational psychology, and information science (i.e., computer science; Smith & Smith, 1994). Cognitive scientists are concerned with research on higher-order abilities involved in thinking and problem-solving. According to Robert Glaser (1990), cognitive scientists are studying the structures and processes of human competence as well as the nature and performance system as consequences of learning and development.

In the past, scientists studied mind or the existing cognitive structure and described how learning occurred. This is how learning is acquired. Today the focus is on the question, "What is learning?" and on teaching how to learn (e.g., communication skills). In the past, learning theorists studied the experience of the phenomenon and how to express it or how to acquire the knowledge from books.

In the present, the focus on learning is different, bending toward how knowledge is organized in the memory, or cognitive structure, and what reinforcement processes are needed to retrieve it. Some reserachers have proposed that knowledge is organized around problems, principles, and abstractions around certain subjects. Thus instruction should be based on providing experiences that will help children organize knowledge in the brain and teach them how to retrieve it by problem-solving and higher-order thinking. This is done by presenting a variety of sensory stimuli to be recognized in parts so that it can be processed, generalized, and stored in the memory in an organized way.

Information Processing

Information Processing learning theories are based on studies of the neurophysiology of the nervous system. The processing and structures reflect the action of the nervous system. They are postulates, as they are not related to particular actions of the brain. Theorists study the structures of the human learner and the kind of processing accomplished by each structure. From these studies, a model of information processing is developed. Gagne (1985) described the process as envisioned by the information processing model, which includes sensory inputs, neurological processing, and performance outputs. Tables 6-1 and 6-2 further explicate this model. The steps are described below.

Sensory Register: A stimulus from the environment enters the sensory register and remains briefly. This sensory stimulus is an image of the real object. It creates attention from the sense organ. For example, the image of a sharp knife is imprinted on the eye or the ring of the telephone reverberates in the ear.

Short-term Memory: The representation is stored in the short-term memory in perceptual form. That is, the image is interpreted by the sense organ in relation to the past experience of the learner. The learner rehearses the information. For example, the learner hears a telephone number through the ear and repeats the number. In this process, old information is pushed out as new information enters the sense organ. As the learner rehearses the information, which then becomes input for the next stage, the long-term memory becomes embedded.

Encoding and Storage in Long-term Memory: Information is transferred from short-term to long-term memory in a process called *encoding*. It is not stored as sounds or shapes but is in conceptual form or meaningful mode. The information is organized as postulates or concepts, paragraphs, or comprehension units such as tables, diagrams, detailed images, or pictures. The main characteristic is that it is organized in meaningful or semantic fashion. The storage in long-term memory is permanent. It may, however, become inaccessible as a result of interference, which blocks memories or ineffective storage, in which case it cannot be retrieved.

Retrieval: Only items that have been learned may be retrieved. To be retrieved, cues must be evoked either by the learner or by external stimuli. The cues recognize the stored information and return it to short-term memory, working memory, or consciousness (Atkinson & Schiffer, 1968). The stimuli may be transformed to the response generator as is the case in motor skills. The recalled memory is reconstructed in an organized pattern or image, in the same form in which they were learned, for example, as with the number of windows in a house. What has been recalled may be applied to a new situation and is called transfer of learning.

The Working Memory: The short-term memory temporarily stores informational acts as working memory. The working memory may initiate processes to retrieve information from the long-term memory. For example, in learning to find the hypotenuse of a right-angled triangle, the working memory retrieves the lengths of sides and combines the information so that the new knowledge may be deduced.

Response Generator: The response generator is the structure that initiates a response to a processed stimulus and determines the pattern of performance.

Performance: Performance activates the effectors, and affects an externally observable behavior such as lifting an arm or screaming. .

Feedback and Reinforcement: Feedback is also stimulated by the environment. The function of feedback is to verify that the act has been performed. Feedback fixes the learning so that it becomes permanently available and is called reinforcement.

Table 6-1. The Information Processing Model

1. Sensory Inputs (the person sees the knife or hears the telephone);
2. Neurological Processing; followed by
3. Performance Outputs (the person screams or takes the telephone).

THE BRAIN	NEUROLOGICAL PROCESSES
Feedback → Reinforcement	
Performance (scream)	
Response Generator	Response Organization
Short-term Working Memory	Retrieval
Search	
Long Term Memory	Long Term Storage
Semantic Encoding	
Short Term Memory	Short Term Storage
Rehearsal	
Perception	
THE EAR Ø» Input » Sensory	Reception of Patterns
Registered	Neural Impulses
Hears the Telephone Ø»	Stimulus Receptor

Adapted from Gagne's *The Conditions of Learning*, p 76.

Table 6-2. The Information Processing Instructioal Model

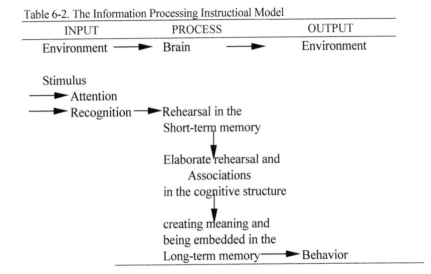

INPUT	PROCESS	OUTPUT
Environment →	Brain →	Environment
Stimulus → Attention → Recognition →	Rehearsal in the Short-term memory ↓ Elaborate rehearsal and Associations in the cognitive structure ↓ creating meaning and being embedded in the Long-term memory →	Behavior

Operationalizing the Information Processing Model

Biehler and Snowman (1997) described how to operationalize the theoretical information processing model. They have transformed it into an instructional model described below. It is as a step-by-step method for examining information processing and learning in a classroom situation. The major steps typically include attending to a stimulus, recognizing it, transforming it into a mental representation, comparing it with information already stored in the memory, assigning meaning to it, and acting on it in some fashion. Information processing the way humans to recognize, transform, store, and retrieve information. Information processing is based on a combination of computer technology and the functioning of the brain. The process consists of the following steps:

1. *Visual or auditory in stimuli enter the brain* from the environment, and the human brain encodes it. The thought process creates a picture of the information and examines the relationships to ideas in the mind.

2. *The brain recognizes the phenomenon and encodes* it via the thought process by noting its features, and then examines the relationships with ideas stored in the brain.

3. *Attention* is focused on the features of the new stimulus.

4. *Information in long-term memory cues the learner as to what requires attention* and checks relevant features and perceptions. Prior experiences direct the individual to see, and in turn what is seen affects what is known.

5. *Short-term memory holds information and rehearses it* with reference to stored information.

6. *Stored material that has been stored in an organized way is recalled, is associated with new material, and creates meaning.*

7. *Cognitive structure or long-term memory* stores new material.

8. *Rehearsal learning becomes permanent* if there is sufficient mastery.

Conclusion

Earliest teachers based their craft on the teachings of the great philosophers. Mental discipline was the earliest learning theory derived from the philosophies of Plato and Aristotle. When science touched philosophy, psychology emerged and gave impetus to a new set of learning theories. With experimental science, psychologists began to conduct experiments first on the sense organs and then on indicators of the senses such as intelligence. Associationist psychology began with the work of Pavlov and Thorndike, which gave rise to the theories of Watson and Skinner called *Behaviorism*. Behaviorism emphasized the interaction between the individual and the environment.

The opposing cognitive theories, Gestalts such as Wertheimer and Field Theorists such as Lewin dealt with how knowledge is configured within the mind and the environment. The synthesis of the two schools was tempered by a new approach to research. This was Cognitive Science and neuroscience, which emerged from research on computer science and information processing. Cognitive science is based on research on the brain by neuroscientists, information science, and psychology. Pedagogy based on brain research seems to hold the key to the future of pedagogy.

ASSIGNMENT

1. Discuss the two types of learning theories and their implications for classroom instruction.

2. Cognitive theories may promote student learning more efficaciously than any other model. Discuss.

2. Describe with examples how the information processing model could be implemented in the classroom.

Chapter 7

The Learning Brain

The human brain is the control center for the whole body. It coordinates with the spinal cord to create the central nervous system, which transmits and receives the stimulation that regulates all mental processes, sensations, actions, and emotions. The brain receives nerve impulses from the sense organs and analyzes those impulses using a process that involves reasoning, intelligence, and associations with past experiences to cause learning. Table 7-1 provides an overview of the process.

Anatomy of the Brain

The Cranium

The human brain is situated in the skull or cranium of the human body, which acts as a protective structure. The arched surface of the cranium and a spongy layer under the skull provide further protection. The cranium is composed of eight bones called *sutures*. The cranial nerves serve the eyes, ears, and nose. The foreman magnum passes through the lower part of the cranium to the spinal cord. Around the brain are the meninges, the dura mater, and the cerebrospinal fluid, which act as protection for the brain. The brain itself is not solid but has cavities or ventricles that are filled with fluid called the cerebrospinal fluid.

Table 7-1. Processing of Information Model

1. STIMULUS
▼▼ ▼
2. SENSORY INPUT
Stimuli from the environment at speeds of 312 to 395 feet per second enter the sensory area of the cortex. Stimuli enter as the neuron nerve impulses and speed from one neuron along the axon and across a gap called a synapse. As the impulses pass over the synapse, the nerve emits a chemical substance that provides the energy for transmitting the signals. For example, a firefighter sees a child running across the street, or the teacher asks a question. ▼
3. SHORT-TERM MEMORY
Sensory impulses travel to the cerebral cortex where they are stored in short-term memory for about 5-20 seconds. Sensory cells travel to the cerebral cortex where they are associated with memory cells that store past experiences. These are interpreted and then relayed to motor cells for action. The firefighter grabs the child, or in the case of the question by the teacher, the child selects the relevant stimulus (i.e., the question) and responds by attending to the teacher. ▼
4. WORKING BRAIN
The neuron selects from entering stimuli. These are processed in working memory (e.g., in discussion, repetition, thinking, and articulating). The child rehearses and interprets the information associating the new stimulus with ideas in memory. Operant responses, such as problem-solving, occur. ▼
5. LONG-TERM MEMORY
The main generalizations are stored in long-term memory including conditioned learning.

The Cranial Nerves

Cranial nerves are either sensory or motor in nature. Sensory nerves are the optic, auditory, and olfactory, which serve the eyes, ears, and nose respectively. The motor nerves are the oculomotor, trochlear, abducens and hypoglossal, which supply the nerves of the neck and larynx and the muscles of the tongue respectively. Other sensory nerves are both sensory and motor. The trigeminal carries sensory impressions to the face, eyes, nose, teeth, gums, and tongue and to the jaws for chewing. Sensory nerves lead from the tongue to the salivary gland and the vagus carries impulses from the larynx and pharynx to the heart, stomach, esophagus, and other organs.

Blood Supply

Energy for the activity of the brain is supplied by blood from two arteries, the internal carotid and the vertebral. The brain consumes up to twenty-five percent of the blood's supply of oxygen.

Structure of the Brain

The human brain has three major divisions: the cerebellum, the cerebrum, and the stem. Each division has a major function. No one division dominates any one function, but indeed all the parts function together as a coordinated whole.

The Cerebellum

The cerebellum is also called the "little brain." It is located at the back of the head under the cerebrum. It has two hemispheres that are joined by a wormlike substance called the vermis. The cerebellum resembles a butterfly. The surface is regular and has parallel grooves with gray matter on the outside and white matter in the center. The function of the cerebellum is muscle activity to is provide feedback. It coordinates movement and balance and maintains posture and voluntary muscle movement. Below the cerebellum is the pons, which contains the nuclei of the sensory nerves and a bulging mass of fibers. The proliferation of fibers and cranial nuclei makes it an important center so that injury to the pons causes coma.

The medulla combines the pons and the spinal cord at the back of the skull into the foramen magnum. It transmits messages from the spinal cord to the brain. The medulla also controls center of breathing, the heart rate, andblood pressure.

The Midbrain

The midbrain is the smallest section of the brain. It contains many nerve cells and fibers, which run from the cerebral cortex, and cranial nerves. The cranial nerve oculomotor supplies the eyes muscles and the clear nerve supplies the muscle that rotates the eye. In the midbrain is also a network of nerves that extend throughout much of the brain to the cerebral cortex. It is called the reticular formation and serves to control electrical impulse of the central nervous system. The reticular formation helps to maintain wakefulness.

Between the midbrain and the cerebrum is the thalamus and below it, the hypothalamus. The thalamus is the seat of consciousness and the hypothalamus controls the sympathetic nervous system. When the anterior hypothalamus is excited, blood pressure and heart rate increase. Stimulation of the posterior hypothalamus causes sleepiness and lethargy. The hypothalamus also regulates the body's water balance, temperature, and metabolism of fats and carbohydrates. Below the hypothalamus is the amygdala, which is the seat of all fears. From it the sympathetic nervous system originates. Below the amygdala is the hippocampus.

The Cerebrum

The cerebrum is the inner part of the brain and the largest segment. It consists of two hemispheres, the right and left, separated by a groove called the longitudinal fissure. Inside the cerebral hemisphere are nerve centers that regulate thought and action. Together the two hemispheres occupy two-thirds of the entire brain and weigh 0.9 grams. Inside each cerebral hemisphere are masses of grey matter called ganglia.

Also inside the cerebrum is a dark grayish matter called the *cerebral cortex*. Beneath the gray matter is white matter, made of millions of cells and called the *corpus callosum*, which is a consolidation of fibers that looks like a cable. The cortical gray matter connects the grey matter with cerebellum, the midbrain, the spinal cord, and areas of the brain.

The nerve fibers from each hemisphere cross each other at the medulla or base of the brain, and then progress down the spinal cord so that each hemisphere controls functions on the opposite side of the body.

The Cerebral Cortex

The cerebral cortex occupies most of the cranial cavity. Only about one-third of the cortex is outside of the cerebrum. The remainder is inside the cortex and consists of grooves and fissures called the fissures of Rolando and Sylvius, which are subdivided into lobes called the frontal, parietal temporal, and occipital lobes. The cerebral cortex consists of nerve fibers, neuroglia, and blood vessels. The axon is part of the cell called the axon which and is one of three types based on function and destination. Projection fibers go to all directions of the nervous system, including the spinal cord. Association fibers are the most numerous types of fibers and make connections with other neurons in the same hemisphere. Commissural fibers connect to the neuron in the opposite hemisphere so that the two hemispheres may communicate with each other.

Cells of the Cerebral Cortex

Within the cerebral cortex are cortical cells that possess dendrites that project vertically and horizontally. There are two types of brain cells: neurons, which are nerve cells, and neuroglia, which are supporting cells. Each neuron consists of a cell body with tiny projections called dendrites that extend to the surface of the cell. There are also extensions called axons that range in length from a fraction to many feet. An example is the axon, which runs from the brain to the big toes. The axons are enclosed in a substance called *myelin* that covers it like a sheath. Myelin consists of proteins and lipids, which look like fibers. These fibers conduct the nerve impulses, conduct the energy for the axons, and act as an insulator. In some parts of the cortex the axons are so densely packed that they have a grey appearance from which they derive the name *gray matter*. In those areas where the axons are less densely packed, the surface is appears white, from which they derive the name *white matter*. The kia are supporting cells that connect all cells into a network. They provide nutrition and are the most numerous cells in the cortex.

How the Brain Functions

The brain operates by transmitting nerve impulses from one cell to another. The impulses travel from the body of a cell along the axon. It transmits across a synapse, which is minute gap between the two cells, and is received by the dendrites of the next cell. These nerve impulses travel at immense speeds of about 312 to 395 feet per second. Electrochemical changes in the surface of the nerve initiate the transmission; the neuron supplies the energy from transmission, which is self-sustaining.

The nervous system has thousands of nerve pathways that lead to and from the brain. The nervous system relays the sensory impulse through the spinal cord to the brain or to cells in the cerebral cortex. From here they are transmitted by the association cells to sensory cells through peripheral nerves to the muscles. The muscle then responds. An example of a typical cycle is the perception of a fire in the basement. Sensory impulses from the eye relay the impression to the cerebral cortex. The impressions are associated with past experience in the memory about a fire. An interpretation is made and impulses are relayed to nerve cells of the cerebral cortex that transmit the impulse to the arms and tells the individual to move. The entire process takes a few seconds (see Figures 7-1 and 7-2).

The Cerebral Cortex

The cortex has three major functions: associative, sensory, and motor, as described below.

Sensory Functions

Inside the cerebral cortex lie the Rolando and Sylvius fissures, which subdivide into lobes. Across the fissure of Rolando in the parietal lobe is the postcentral gyrus. This is the main area for sensations and is called the *somesthetic area*, which receives sensations of touch and taste. The temporal lobe receives hearing sensations, vision is located in the occipital lobe, and smell in the temporal lobe. The thalamus receives pain and temperature signals.

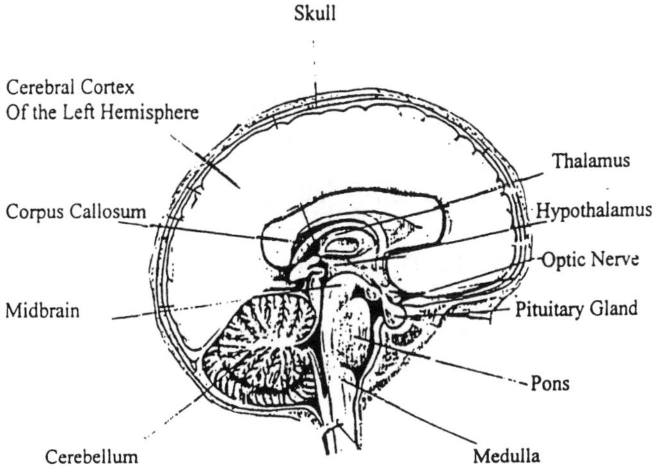

Figure 7-1. Cross Section of the Human Brain

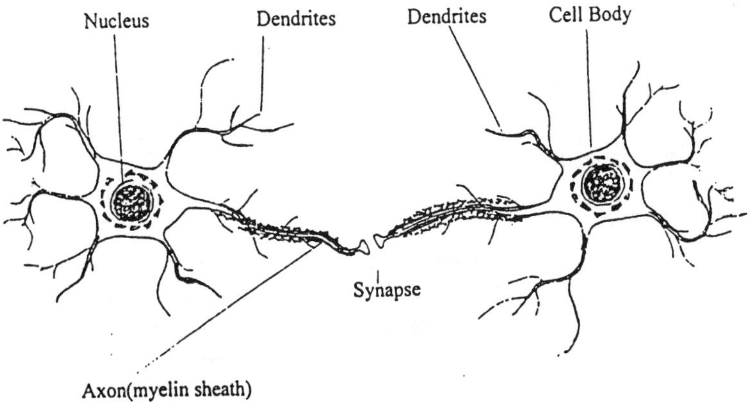

Figure 7-2. Brain Neurons

Motor Functions

On the frontal lobe of the Rolando is a long body of tissue. On this lobe is a ridge called the *motor cortex*, which can be mapped by electrical stimulation. The motor area is responsible for the jaws, tongue, and throat. Between these areas are the centers for the hip, trunk, shoulder, hands, and face. The cells in this motor area are called the *Betz cells*. The axons of these cells make up the corticospinal tracts that are shaped like a pyramid and descend into the midbrain and into the spinal cord.

Memory: The Associative Function of the Cerebral Cortex

The mass of associative cells of the cerebral cortex is called *memory*. Associative cells are the most numerous in the cortex. These are linked together in chains by associative fibers. These cells are reused constantly, and the greater the use, the greater ease with which the synapses become useful.

Memories stored within these cells create a mass that makes it possible to compare new impressions with previous ones and to reach conclusions. Memories receive stimulation from new sensations that are relayed by parasensory areas that combine them into more elaborate impressions capable of being recalled. Sensory areas have locations as well as form, size, and texture. This enables new sensations to be recognized and matched. Most memories are stored in a complex pattern in the temporal lobe, although some are stored elsewhere. The prefrontal lobe receives associations from the thalamus, which is concerned with emotions. Removal of the thalamus results in a removal of emotions as in patients who have had lobotomies.

Phases of Memory

Memory has two phases: a short term and long term. In short term memory, the brain receives impulses which last up to an hour, and then become extinguished if they are not reinforced. Short term memory may be eased by electrical shock or injury to the brain. This causes the memory to be lost. However their effects persist and may be recalled under hypnosis. Hypnosis may allow a person to recall memories which have been lost for years, and even memories from childhood.

Storage and Recall of Memory

Memories are stored when permanent changes occur in the synapse. The theory is that there are changes in the ribonucleic acid molecule (RNA). Researchers believe that memories are stored in the coded proteins of the cell.

The Learning Process

Learning is a synthesis of present experience and present event. It occurs because of the functioning of the brain in three processes:

1. Recall past experiences and impressions in memory; and
2. Association, which is the ability of the brain to associate past and present impressions and synthesize them into a new creation; and

Learning may be described either as conditioning or cognitive functioning.

Conditioning

There are two types of conditioning: classical conditioning, developed by Ivan Pavlov, and operant conditioning described by B. F. Skinner. Classical conditioning is a conditioned response to paired stimuli or two stimuli provided simultaneously in a situation. The response is reinforced by rewards. Learning is routine. Operant conditioning, on the other hand, requires that the organism first emits the behavior. The behavior is rewarded. Rewarded behavior is reinforced, but this only happens when the subject perceives the behavior to be purposeful. This leads to cognitive learning and creative thinking.

Cognitive Functioning

Cognition assumes that the brain interprets new information. This occurs when new stimuli are associated with past experiences. The resulting behavior is stimulated by mental capacity, needs, past experiences, motivation, concentration, emotion, attitudes, and values.

Inherent Characteristics of the Brain in Learning

Learning is a cognitive process that occurs in the cortex of the cerebrum. It involves the interaction of stimuli from the environment with memories in the cerebral cortex, along with other operating functions of the brain such as intelligence and emotions; and factors in the environment of the brain such as the electrical activity, chemicals, nutrition, oxygen supply, and consciousness.

Intelligence

Intelligence is the gene in the deoxyribonucleic acid (DNA) and the ribonucleic acid (RNA) that forms the chemical substance the body needs. In the process of growth or when the cell comes in contact with stimuli, it divides itself, and a section of the DNA ladder splits apart to direct bodily function such as the creation of a chemical that the body needs for self-repair. By this method it creates a chemical substance called RNA, a messenger that carries instructions to the new situation, in this case incoming nerve impulses from an encounter with a stimulus in the environment. Scientists believe that learning takes place when new stimuli cause changes in the cells' DNA, and because RNA plays an important part in the synthesis of proteins by the cells, it is believed that memories are stored in the coded proteins within the cell.

The Environment and Learning

Biochemistry affects the brain and hence learning. Medical researchers have discovered that physical changes such as glandular disorders can be triggered by chemical deficiency. Glandular changes can be caused by thyroid deficiency, and coma or convulsion can triggered by administration of certain drugs. The discovery led to analysis of chemical substances of the body that can be done in any laboratory and the following discoveries. Brain cells consist of proteins, lipids, and carbohydrates, all of which are present in foods. In all, ninety percent of the brain consists of protein or amino acids. Researchers have shown that amino acids influence the functioning of the brain. For example, nylalanine causes mental deficiency. Glutamic acid is involved in glucose metabolism and acid may be used in the treatment of mental problems such as epilepsy, schizophrenia, and alcoholism.

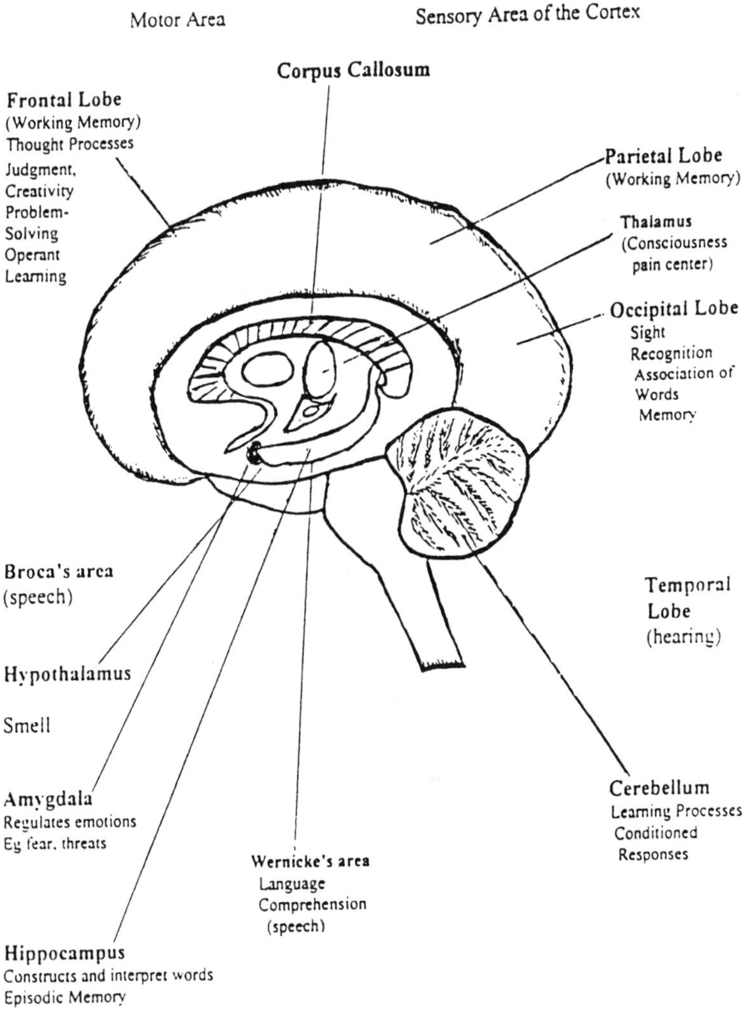

Figure 7-3. The Cerebral Cortex: Sensory Area of the Brain: Association, Memory

Tryptophan produces niacin and serotonin. Niacin is related to pellagra and serotonin is related to sleep and to the use of tranquilizers to induce sleep. Psychedelic drugs contain chemicals that influence brain function. Drugs such as LSD cause hallucination. The implication is for education of students in prevention drug use.

The brain gets its energy from blood. Although the brain comprises only two percent of the weight of the body, it consumes twenty percent of the body's energy. To function properly the brain must be fully hydrated. It needs eight to twelve glasses of water per day to keep it in good condition. Hanaford (1995) showed that dehydration leads to impaired learning. Oxygen is also necessary for the brain. Concentration causes to consume more oxygen; oxygen increases mental alertness.

Educators must inform students about proper dietary habits such as a protein-rich diet to enable the brain to function efficiently. Ostrander and Schroeder (1991) and Hutchinson (1994) discovered that vitamins and minerals can boost memory, intelligence, and learning. Howard (1994) found that calapin dissolves protein build-up and cleanses the synapses and thus transmits nerve impulses more effectively. Calapin is found in yogurt and leafy greens, especially spinach and kale.

Learning is a product of intelligence and environmental activity, with intelligence occupying thirty to sixty percent and environment forty to seventy percent. Thus educators must concentrate on nurture in the environment. Diamond discovered enriching the environment produces brains with thicker cortexes, more dendrite branching, more growth spines, and larger cell bodies. In effect stimulation enhances the learning process. Some researchers have found that smart people have more neural networks, especially with learning and memory.

Dendrite branching depends on the type of activity in which the brain is engaged. Novel experiences generate new synapses, and exercise creates more layers of motor molecules. Hence new experiences act as challengers to the brain and repetition creates more learning. The implications for learning are to give the brain novel and challenging exercises. New materials, new levels of difficulty, and new strategies will all interest and help the brain to grow and learn more effectively.

Emotions and Learning

During the twentieth century many studies have been done on the effect of electrical charges on the brain. German scientist Edward Hitzig and Gustav Frisch discovered in 1870 that a subject made specific motor and sensory responses upon electrical stimulation of the cortex.

The original studies were done on animals and later on human subjects. Although there were several inconsistencies, the basic result was that certain muscles responded when specific areas of the brain were stimulated. This course of study was pursued in the twentieth century with more sophisticated electronic equipment such as Magnetic Resonance Imaging (MRI) and PET and CAT scans. This led to the discovery of the electroencephalogram (EEG), which facilitated the discovery that if ceratin areas of certain are stimulated, sensory and motor responses result. When the thalamus was stimulated on one occasion, the subject fell asleep; another time the subject was aroused. The conclusion was that stimulation of the thalamus produced emotion including pleasure, pain, and rage.

In essence the brain reacts to external influences. Negative influences such as threats, embarrassment, unrealistic deadlines, humiliation, sarcasm, lack of resources, and bullying have negative effects and should be removed from the environment (Jensen, 1998). LeDoux (1996) found that the amygdala is the center of all fears. From it extends a systematic nervous system, that is, neural project bundles of fibers that trigger the release of adreneline, vasopressin, and cortisol that immediately change the way we think, feel, and act. In a classroom situation, therefore, threats should be avoided at all costs.

Another negative influence is stress. Vincent (1990) showed that stress causes the adrenal glands to release peptides or cortisol, which occurs when the body faces physical, emotional, or academic danger. The resulting physical reactions are depressed immune system, hypertension, and blood clotting. Chronic stress therefore leads to greater susceptibility to illness and disease, which can influence academic performance and grades. Educators should be realistic in the demands for academic exercise so as not to create stress in students. Most important, stress affects memory. Vincent (1990) found that chronic stress leads to the death of brain cells in the hippocampus, influencing memory formation. The implication is that the school should pace the rate of the students' world so as to avoid creating stress for students. The implication is that emotions operate in learning and school discipline.

Electrical Impulses

In 1875, Richard Caton, an English physician, studied the brains of various animals with a galvanometer and found that the brain is always active, even in sleep. It constantly sends and receives nerve impulses or electrical activity. Later, in 1929, a German neurologist recorded this electrical activity on paper as a wavy pattern called *brain waves* that are measured in amplitudes or height and frequency or number of waves per second. Electricity or brain impulses carry stimuli from the source of interaction to the brain.

In 1929, Hans Berger discovered the presence of alpha rhythm, which seems to be associated with the eyes. Alpha rhythms are most prominent in the occipital part of the brain during times of concentration and are interrupted when this activity ceases. Electricity or nerve impulses therefore are most active during times of concentration; the strength of the stimuli accelerates the speed of electrical impulses and increases learning. There are also beta and delta waves. Electrical activities are affected by chemical and physiological changes in the body. In learning, environmental stress levels have to be controlled.

Brain Chemistry

The brain consists of ninety percent protein, which is necessary in its metabolism. Because new knowledge is encoded in proteins, there must be a supply for protein for effective learning. In addition nutrients such as vitamins and minerals operate in the oxidation of proteins and must therefore affect the quality of learning.

Consciousness

Neurophysiologists and neurochemists have not yet arrived at a conclusion about the nature of consciousness. Nevertheless some theorize that consciousness is in the midbrain, as destruction of the cortex deprives an individual of memory but not of consciousness. The individual can still respond to pain in the environment. The center for pain, however, is the thalamus, which acts as a relay station for impulses coming into the brain. In learning, stimuli that create pleasure are more likely to increase the response to the environment and so to promote learning. A conscious act is a decision based on previous experience. Hence consciousness involves the interaction of the cerebral cortex, the brain stem, and the thalamus.

Memory or Associations

Memory, the sum of past experiences is brought to consciousness by the brain and interacts with incoming nerve impulses and the chemical environment of the brain and the intelligence in the DNA to create learning. Emotion which is stimulated by the thalamus and also involved in learning.

Environmental Stimuli and Attention

In learning, each brain cell acts like a tiny electric battery. The cell is energized by the minerals sodium and potassium. These act as neurotransmitters and provide energy for dendrite growth as they touch other cells. The cell body sends an electric charge through the axon and releases the stored chemicals into the synapse. The chemical reaction triggers an electrical response in the dendrites. The process is electrical to chemical and back to chemical. The process repeats from cell to cell. The process results in dendrite branching. New synapses are an indicator that learning has taken place. Stimulius arrive in the brain through the senses. The quality and strength of the stimuli are important predictors of learning. These must be attention- getting. At any one time there are numerous stimuli in the environment, but the brain can only focus effectively on one at a time. The brain selects the stimuli through the senses, especially the eye, which handles eighty percent of the images thatenter the brain. The information flow is a two-way process from the environment to the eye, to the thalamus, to the visual cortex.

This feedback mechanism shapes attention so that the individual can focus on one thing at a time, such as the lesson being taught. There are several images getting feedback at any one time. The volume triggers selective neurons along the visual pathways to fire less often because their membranes are hyperpolarized. This prevents normal processing. In effect the brain selects some stimuli and ignores others.

Attention is conditioned; individuals select something because they are told to do so. Attention is therefore selecting some feedback data and excluding others.

Hobson (1994) suggested that chemicals have much to do with what the brain selects. Chemicals such as neurotransmitters, hormones, and peptides are involved. Acetylcholine, a neurotransmitter, is linked to drowsiness; adrenaline is linked to alertness; and norepinephrine is an indicator of attentiveness. Low levels indicate inattentiveness, and high levels indicate hyperactivity. Because a change in chemicals can cause a change in behaviors, the teacher should schedule activities to accelerate chemical reactions that facilitate learning. Options include the use of creative and challenging activities, physical activities, or any attention-grabbing stimuli to evoke attention.

Educational Implications

The role of the teacher is to provide the enriched learning environment in which learning can occur. The quality of the environment is a motivating factor in learning. The teacher also should educate the students in the nutrient requirements and drugs to avoid in the learning situation.

The teacher should employ the process that will facilitate learning such as conditioning, evoking past experiences, and reinforcement. Organization of content is very important because content is encoded in proteins. New facts should be presented in an organized way that will facilitate the process of encoding in the memory, which also facilitates recall of past experiences.

ASSIGNMENT

Create four groups and discuss the following topics.

1. Imagine that a girl sees an oncoming train. The girl screams. Describe the cognitive processes that take place during the entire period of time.

2. How do emotions affect the behaviors of students in a classroom?

3. Describe the influences of the environment on learning.

4. What influence do past experiences have on learning?

5. How does nutrition impact on student learning?

6. Describe how the sensory organs function in learning.

7. Describe how you would use knowledge of the functioning of the brain in the teaching-learning process.

Part 3

Integrating Theory into Practice

Chapter 8

Approaches to Teaching

"Often teaching and learning are thought to be correlative terms like selling and buying, one term implies the other" (Brubacher, 1969, p. 235). It is not consistently factual that where there is a learner there is a teacher; the fact is that there are those highly motivated individuals who can teach themselves. It is generally acknowledged, however, that a teacher facilitates the learner. To be an instructor, the teacher draws on subject matter knowledge to decide on content and on training in epistemology to determine the general approach and specific methodology. Teaching and learning embrace each other.

The approaches to teaching that are described in this chapter are based on the theories discussed in the preceding chapters. Pedagogical theories representing the aims of education, the curriculum, and the teaching-learning process are based on philosophy, particularly epistemology or the theory of knowledge. Most theories embrace a dualistic conception of mind, corresponding to a dichotomy of aims of education, and methodologies from authoritarian to democratic teaching-learning process.

Psychological theories of learning similarly have a dualistic conception and range from stimulus/response-associationist and behaviorism to cognitive-field and developmental and neuroscience. These learning theories will be applied to the teaching-learning process in the classroom situation. Finally, knowledge of how the brain functions and theories of information processing will explain how the mind learns, stores and retrieves information.

Approaches Based on Epistemology

Generally, philosophical beliefs, and psychological principles, give rise to a myriad of pedagogical principles and methodologies. Nonetheless, teaching methodologies may be subsumed under two main approaches to teaching. At one extreme is the authoritative method of exposition and at the other is the problem-solving method and method of inquiry, with any number of approaches appearing on a continuum in between.

Methods of Authoritative Exposition

Also called traditional, the method of authoritative exposition has been practiced by educators since the times of the Greeks in the first millennium B.C.E. This method was based on the learning theory of mental discipline that mind is some invisible entity within the learner. Plato conceived of knowledge as emanating from innate ideas and learning was by reasoning. His theory was based on the premise that empirical knowledge gained through the senses was illusionary and inconstant and therefore untrustworthy. Aristotle proposed that both mind and matter are real and that learning was by experiences and reason. Curiously enough, these points of view exist even today as educators think of the rationalist as one who sits on an armchair and reasons while the empiricist goes to a laboratory to make observations. Both Plato and Aristotle were concerned with the concept of mind and hence arrived at the same conclusion: the aim of education is mental discipline, that is, to fill the mind with ideas.

Rationalism as a theory was a refinement of the Platonic Idealism. Descartes, Spinoza, and Leibniz reduced the concept of innate ideas to a few basic principles from which truth may be deduced by logic. Descartes said that knowledge may be derived from a few principles or axioms by applying deductive logic. As an opposing idea, John Locke (1632–1704) refuted the concept of innate ideas and promoted the "modern way of thinking, defined idea as the immediate datum which mind experiences when it perceives an object, members, or understands." When one perceives a tree, the tree does not enter the mind, only the idea of the tree. Locke's theory was a refinement of Aristotlean Realism. Learning was derived from experiences, the datum and reasoning about it. These theories were the basis of the pedagogy of mental discipline, this in effect means mind plus discipline or subject.

The theory of *Mental Discipline* as an aim in education, begun by Plato and Aristotle, continued to dominate educational epistemology for centuries and may be evidenced in today's classrooms. The traditional use of the authoritarian approach to teaching is central to this epistemology. Based on philosophical analysis, authoritarian teaching is derived from metaphysics, epistemology, and axiological aspects of philosophy, and its application to pedagogy. The authoritarian approach of pedagogy is basic to mental discipline, a learning theory in which education is for the development of the mind and the self of the pupil, and activities are intellectual.

By definition, pedagogy involves the subject matter content, the teaching-learning process, the teacher's leadership style, and methods of evaluation. Each component will be different based on the approach to teaching. The subject matter content under the authoritarianism approach to teaching is subject centered. It is derived from the philosophies of Idealism and Scholasticism and parts of Realism. These philosophies are based on the idea that knowledge is absolute and complete. The implication is that subjects are constant and essential to the realization of the pupil's potential. Constant knowledge is of the past, the cultural heritage of the race, or the best experiences of human culture. Some expressions of the cultural heritage are art, biography, history, literature, music, and especially mathematics.

Whitehead (1962) called a curriculum that is rooted in the past "inert ideas" that are received into the mind but without being utilized, tested, or thrown into fresh combinations. The only purpose of inert ideas is to equip the learner for the present. Whitehead believes that schools are overladen with inert ideas and that propositions are used in isolation. An example of such inert is knowledge is in mathematics when it is taught as a cultural subject with theories, logical analysis, and proofs. Because such knowledge is inert, it is not applicable to the solution of current problems.

Teaching-learning processes that are subsumed under authoritarian methodologies are lecture and the method of exposition. These are exemplified by Herbart's five-step instructional model. In the authoritarian methodology, the organization of the teaching-learning process follows the logical order of the subject content. That is, the lesson must be organized into a sequence of interacting parts that are taught by stages. Authoritarian methodology requires the teacher to be an expert in the subject matter content.

In both Idealism and Realism, the teacher's leadership style is authoritarian. This means that the teacher makes all major decisions in the classroom and everyone else concurs. The lesson involves a teaching-learning process in which the students are required to repeat and remember facts. The teacher then leads the children to reflect on what has been learned and to derive meaning.

Because of the authoritarian leadership style, the discipline of the classroom is dominated by the teacher's personality. Theories of leadership and style may be based on the Need Theory of motivation proposed by McClelland (1968). These are the Need for Power (N-Power), the Need for Achievement (N-Ach), and the need for Affiliation (N-Aff). In the authoritarian classroom, the teacher is the legitimate, referent, and expert authority, hence the term authoritarian. The teacher demands attention, initiates activity by making decisions on what should be accomplished, and gives rewards for effort and achievement by symbols such as grades.

Exposition

Expository methods and lectures are the major methods used in authoritarian classrooms. This approach assumes that one person is an expert in possession of authenticated knowledge that another lacks and desires. Authoritarian relationships include a teacher and pupil, lawyer and jury, or president and board of directors. The teacher is able to control the direction the pupils logically and step by step to move toward the goal of the lesson. The method is similar to that of a teaching machine that directs the learner through a series of steps of a scope of knowledge. The teacher's skill is communication, and from this perch the teacher pontificates truth from a high moral ground while the students submits to his direction.

The knowledge that the teacher conveys is that which has been authenticated by the race, the cultural heritage, and facts from the past, Beginning learners would know very little of the historical culture and could not logically discover this for themselves. The method is to use deductive reasoning to gain this knowledge. The methodology of authoritarian exposition is therefore very mechanical. The learning of the cultural heritage assumes that the aim of education is fixed by the social heritage. Such assumption is consistent with the metaphysical stance that knowledge is constant, immutable, and eternal.

In re-creating the past, the mind is passive; that is, it absorbs the material much like a sponge absorbs water. Mind is not active in that it not using cognitive functions. What the teacher is doing is building a structure within the pupil, that is, building or filling the mind with ideas or mental discipline.

The content-based, authoritarian building of structures or ideas with the mind lends itself to the method of lecture or the Herbartian five-step instructional model.

Herbart's Instructional Model

Herbart was both a philosopher and psychologist. Herbart held that mental contents, without exception, arise from experience, when primeval consciousness interacts with sensations. Mental contents are presentations that arise form patterns of association. Whenever mind acquires new knowledge, it is assimilated, and the process of assimilating new knowledge to old is called *apperception*. Ideas are able to rise to consciousness and to be associated and bonded with new sensations. The more numerous the associations, the stronger will be the bonds, and the greater the ease with which the ideas are recalled and remembered.

Herbart proposed a five-step instructional pedagogy in his *Outline of Educational Doctrine*. The steps are:

1. *Preparation*: To help the child to recall familiar ideas that will make the new idea assimilable.

2. *Presentation*: Of the new material.

3. *Association*: Enables the mind to connect past and new learnings. The teacher directs the process of making association or comparison and contrast between the old and new ideas, so that in mind these develop associative bonds. The new ideas are assimilated. Examples of association are comparison and contrast, cause and effect, antecedent and consequence.

4. *Generalization*: The process of deriving or summarizing the main points and stating the concept name, rule, or principle. The teacher then makes

generalization by presenting new instances which are based on the same principles as the new material. For example, if the teacher deals with democracy in America, he could compare it with democracy in other countries such as Greece.

5. *Application*: Occurs when the teacher gives the child an assignment or a test in which the students must apply the rule.

Herbatian five step method of exposition has been the most widely used method of teaching in schools over the years.

Nonauthoritarian Approaches

Philosophy also gave rise to nonauthoritarian methods. This approach, also called *problem method of inquiry*, is based on the philosophy of Pragmatism. Because the teacher already knows the answer and the problem only relates to the student, the problem method may lead to authoritarianism.

Problem-Reasoning Model

Problem-solving methods were first introduced by John Dewey (1896–1952). He was influenced by the evolutionary theory of Darwin and by the pedagogy of Rousseau and John Locke. In his books *Democracy and Education* and *Experience and Education*, Dewey claimed that knowledge is integral to the self-sufficient being as part of the process by which life is sustained and evolved. Dewey termed the senses "gateways of knowing" and indicated that they ultimately serve to stimulate action. Dewey suggested that man adapts to the environment by solving problems. Problems are based on life rather than on school subjects.

The problem method is based on the philosophy of Pragmatism by William James and Instrumentalism by John Dewey. Dewey proposed the "the problem should not start with some school subject like mathematics, but rather with some life experience." The next step is to define the perplexity of the problem and then to search for solutions. The fourth step is that the children will construct their own hypothesis, and the fifth stage is to test the solution. The intent is to teach the children *how* to think, rather than *what* to think. The material should therefore relate to the present.

The curriculum of the school should be centered around projects, both individual and group-oriented. In application, problem-solving methods should lead to joint investigation by teacher and pupils. Knowledge will be the outcome of investigation. In this regard, a school that employs this method will require a good library and laboratories. They will focus on activities such as field excursions and displays. Thinking is an outcome of the teaching-learning process; the teacher should facilitate the students rather than regard the learner as an object. The pupils will learn to be independent thinkers.

The problem-solving method can hardly be used exclusively. It has to be supplemented by exposition. This makes the methods of exposition and problem-solving complementary to each other rather than antagonistic; each has a place as an approach to teaching.

Teaching Approaches Based on Psychology

Psychology has given rise to teaching opposing methodologies from Behaviorism at the extreme right to Cognitive theories on the left. On one extreme, the learning theory of Behaviorism gave rise to its own teaching method of direct instruction, which is basically conditioning of students. Behaviorism is similar in procedure to exposition but excludes the concept of mind.

Behaviorists discredit the traditional idea of mind, which is thought to be invisible and therefore nonexistent. The idea of mind has been replaced by that of "observable behavior." The theory is based on the conclusions of the physical changes that are observable, that is, the change in the learner's behavior.

On the other extreme of the continuum are cognitive theories that deal with the processes inside the brain rather than external observable behaviors. These include concept development, concrete representation, discovery, problem-solving approaches, and generally constructivist approaches.

Behaviorism

Behavior Modification is derived from the theory of Behaviorism. Learning is considered the accumulation of new behaviors through conditioning and reinforcement. There is a relationship between the behavior and the environment, the latter providing the stimulus that makes the subject emit responses. Behaviorism began at the time when science touched philosophy and gave rise to psychology. It was based on the Associationist theories of Pavlov and Thorndike. These inspired Watson and B. F Skinner to develop their theory of Behaviorism as discussed in the previous chapter. Behaviorists were more interested in the observable behavior of the subject rather than in mental functions.

In Operant Conditioning proposed by Skinner, the subject first emits a behavior that is rewarded and reinforced by the teacher so that it will have a probability of recurring. The principles of instruction consist of the following stages:

1. *Selecting the stimulus*: The teacher selects the stimulus, which is verbal in nature.

2. *Providing operant reinforcement*: Reinforcement is given immediately the student emits a response. Natural reinforcers are those that already exist in the classroom (e.g., finding the right word to describe something).

3. *Resolving temporary confusion*: Advance to the next stage of the event. There is the application of contrived reinforcers such as verbal comments, gold stars, early dismissal, and free time. Adverse stimuli should be avoided (e.g., ridicule, criticism, detention, trips to the principal's office).

4. *Teaching-learning process*: Teaching the content of the lesson involves shaping behaviors. The subject matter content is divided into sequence and presented with reinforcers at each step. According to Skinner (1954), complex repertoires may thus be developed.

5. *Shaping verbal behaviors*: Identifying the suitable verbal behavior. The teacher then continues with verbal sequences called frames which is in sequence on a scale from no knowledge to competence.

Behavior modification is conditioning and may be used effectively in the teaching-learning process.

Direct Instruction

Blair (1998) defined *direct instruction* as "explicit, systematic, or structured leaning" (p. 51). It specifies a goal in terms of observable student behavior and quantifiable competency level. Direct learning can be broken down into a series of parts; the teacher explains the task at hand in small steps using examples and counterexamples. The deductive process is generally used. It begins with the teacher informing students of the objective of the lesson and giving a definition. The teacher then uses examples to support the knowledge at each step. The next step is teacher-supported practice in which the teacher works examples within the class. This is followed by independent practice, which is usually seatwork or homework. Rosenshine (1986) delineated six instructional functions.

1. Orientation: Check homework first, which is done at every lesson. Reteach as necessary. Review past learning and prerequisite skills.

2. Presentation: The teacher states the objective of the lesson and writes the explicit objective on the board. The teacher explains the content in small steps at a rapid pace. At each stage, the teacher checks for understanding and highlights main points. The teacher gives examples and models acceptable behaviors.

3. Guided practice: Students practice the new concept under the guidance of the teacher. Teacher asks questions to ensure clarity, checks for understanding, and gives feedback. Students continue practice until they understand. Teacher makes corrections and gives feedback. Teacher monitors for errors, gives the right answers, and rewards with praise.

4. Independent practice: Seatwork. Students practice until their responses are firm, quick, and automatic. The teacher monitors students.

5. Weekly and monthly review: Teacher reviews homework and gives frequent tests. Errors on tests prompt reteaching of material.

According to Blair (1998), direct instruction involves the teacher's control of the class. The teacher directs the learning, controls the pace of the lesson, decides what instructional activities will be used, monitors students' progress, decides when the objectives have been attained, and evaluates performance. There is little noise in the classroom and no social interaction between students, only between the teacher and student. There is no movement. The questions used are convergent in nature.

The main activities of the teacher are to explain, inform, and model or demonstrate what the children should know. The heart of the teaching is the manner in which the teacher proceeds to teach the subject with a step-by-step explanation, and the lesson is broken down into meaningful units. Feedback is given at each step, including rewards. This is conditioning and is consistent with behavior modification.

Cognitive Theories

For years, cognitive theorists competed with behaviorists for recognition and challenged them on what happens between the stimulus and the response. Whereas behaviorists were concerned with the interaction between the individual and the environment, cognitive theorists were concerned with how the individual perceived the environment. They believed that how the individual configured or organized the environment was more important than just responding to the external forces. These configuration theorists were called Gestalts and field theories and were influenced by the works of Rousseau, Locke, Pestalozzi, and Froebel, who were empiricists. Cognitive theorists were also influenced by the Functionalism of John Dewey as discussed previously. In later years, cognitive theories were subsumed by theories of information processing and cognitive science, which were based on the functioning of the brain. Cognitive theories include concept development, inductive reasoning, problem-solving, and discovery.

Concept Development

Concept development involves discovery of relationships among variables and making generalizations. There are three phases:

1. *Presentation* of data that defines the concept. Students must identify and compare the criterial attributes of the concept by citing exemplars and nonexemplars. Based on these examples, students generate and test hypotheses.

2. *Identification of attributes* by students for other examples and confirming attributes.

3. *Generalization* involves students summarizing main attributes and making generalizations, then generating other examples.

In essence, concept development begins with concrete examples and ends with abstract definition. Concept development is synonymous with the Method of Discovery.

Inductive Reasoning

Inductive reasoning also called the *Socratic Method,* which is the method of questioning. It consists of the following strategies:

1. *Identifying a problem* and stating it in the form of a question or hypothesis.

2. *Forming a concept,* including identifying concrete examples, enumerating the data relevant to the topic, grouping data by identifying common properties with common attributes, and establishing names for the attributes.

3. *Interpreting data,* including identifying and exploring relationships, such as similarities and differences, cause and effect, and antecedent and consequence, and then making inferences.

4. *Applying principles,* which means predicting consequences from the relationships and then supporting predictions with data and verifying the prediction.

The basis of this strategy is activity and questioning. The essence is analyzing relationships and making inferences and predictions.

Cognitive Science

During the last two decades of the twentieth century, two events influenced learning theories. These are advances in computer technology and information processing and studies in neuroscience. Cognitive psychologists and neuroscientists conceived of research on the brain as an information-processing entity. The synthesis of information processing and brain functioning have evolved into the field of cognitive science. Cognitive science conceives of pedagogy more as learning than as teaching. In the past the emphasis of the teacher was on the acquisition of knowledge, that is, on how learning talks place. For example, the teacher concentrated on the communication skills needed to read a passage rather than on what the student understood. Cognitive scientists focus on what learning takes place (Smith & Smith, 1994), that is, on the organized knowledge that is imprinted on the brain as a result of the learning experiences.

Glaser (1990) suggested that whereas in the past researchers studied the abilities involved in problem-solving, today cognitive scientists study "the structures and processes of human competence and the nature of the performance system as a consequence of learning and development" (p. 29). In essence the emphasis is on meaning that the student has derived from the learning experiences.

Discussion on the functioning of the brain in a previous chapter described how memory is organized and the processes needed to solve problems. Based on this knowledge a teacher may organize instruction on how the brain functions by emphasizing experiences and higher-order Thinking, including problem-solving. Cognitive skills include presenting knowledge using several senses simultaneously, presenting knowledge sequentially as in information processing, using images in symbolic forms, using memory devices, correlating emotions with the learning experiences, and associating ideas. The following instructional models are based on these theories.

Gagne's Conditions of Learning

Robert Gagne (1985) stated that learning is indeed complex and multifaceted. It is the mechanism by which an individual becomes a functioning member of society. Learning is the process by which all of the skills, knowledge, attitudes, and values that are acquired by human beings are assimilated. Learning therefore results in a variety of behaviors that he called *capabilities*. Capabilities are outcomes of learning. Gagne claimed that he was disillusioned with the grand theories of learning. He therefore created a new instructional model. He claimed that successful instruction was consistent with of a set of components with the following sequence:

1. Providing instruction on the set of components tasks that build toward the final task,
2. Ensuring that each component is mastered, and
3. Sequencing the component tasks to ensure optimal transfer to the final task (Gagne, 1985).

Gagne identified five varieties or categories of learning. These are verbal information, intellectual skills, motor skills, attitudes, and cognitive strategies, which are all learned in different ways and are called internal conditions.

1. *Verbal information* is retrieval of stored information such as facts, labels, and discourse. Performance of this capability is by stating or communicating the information in some way, as in stating a definition.

2. *Intellectual skills* are mental operations that permit individuals to respond to conceptualizations of the environment. Performance involves interacting with symbols in the environment, such as the area of a triangle.

3. *Cognitive strategies* are the capabilities that acts as a control panel that governs the learner's thinking. Performance includes managing remembering, thinking, and learning, such as writing a term paper.

Based on these definitions Gagne identified nine stages that must be followed in sequential order that are essential to learning, as shown in Table 8-1.

Problem-solving-Inquiry: From Facts to Theories

This strategy was developed by Suchman who believes that students can be taught to generate theories. The Problem-solving Model begins with a realistic problem. Students are puzzled and develop cognitive disequilibrium. The next stages are data collection and verification.

1. Confrontation with a Problem situation which creates cognitive dissonance
2. The teacher and students analyze the problem identifying the questions raised and the data needed to solve the problem. This data will be the facts to be learned. Students will construct hypothesis.

3. Isolate the variables and the relationships
4. Gather data and give experiment and give explanations
5. Organize data, formulate explanations
6. Verify and analyze strategies.
7. Make generalizations and formulate rules, and explanations

Table 8-1. Gagne's Nine Essential Stages of Learning

DESCRIPTION	LEARNING PROCESS	INSTRUCTIONAL EVENTS
Preparation for	▼ Attending	Alerts the learner to the stimulus
	▼ Expectancy	Orients the learner to goal
	▼ Retrieval	Permits recall of relevant pre-requisite information capabilities in the working memory
Acquisition	▼ Selective	Presenting the stimulus Perception of important stimulus Temporary storage of material features in working memory
	▼ Semantic Encoding	Transfers stimulus features related information to Long-term memory
	▼ Retrieval	Retrieves stored information to the individual's response generator and activates a response
	▼ Reinforcement	Confirms learner's expectancy of learning goal Provides feedback Assesses performance
Transfer of Learning	▼ Cueing-Retrieval	Provides cues for recall
	▼ Generalization	Enhances transfer of learning to new situations

Source: Adapted from Gagne's Conditions of Learning

Conclusion

Professional teachers are committed to their craft. The basic approaches are based on philosophy and have been described on a continuum from authoritarian to nonauthoritarian. Exposition and direct instruction strategies may be described as authoritarianism and problem reasoning is nonauthoritarian. Approaches based on psychology range from behaviorism to cognitive field and neuroscience. Behaviorist strategies include behavior modification and cognitive strategies include inductive reasoning and concept development. Cognitive science involves information processing. The chapter describes some of these models. A competent teacher may use and adapt these strategies to classroom processes.

ASSIGNMENT

1. Write a plan for teaching a lesson by direct instruction

2. In your working groups, identify three models of teaching and discuss how you would implement them.

3. Discuss cognitive theories of learning.

4. Describe the process of learning through cognitive science.

5. Compare authoritarian and nonauthoritarian methods of instruction.

6. Devise a model for inquiry-discovery method of teaching.

Chapter 9

Principles of Planning for Instruction

Teaching does not "just happen." Even those who claim to be natural-born teachers have to plan. The skill with which the teacher plans affects the effectiveness of the instructional process, the success of the teacher, and the learning of the students. Instructional planning has been described by some researchers as the process by which teachers select among alternatives, teaching materials and methods as well evaluative techniques. The purpose of this process is to choose among alternatives that will contribute to a lesson that is focused and effective.

Educators are guided by philosophies of education from which are derived principles of teaching. These principles are then translated into methodologies for presenting knowledge in the classroom. Current educa-tional principles include real-life situations, experiential background, motivation, attention and interest, activities, and use of resources.

The Real-life Context

Every principle in mathematics has an empirical base. Indeed the invention of mathematics was for the purpose of solving real-life problems. It was only later, especially when it was returned to Europe, that mathematics became more abstract. At that time it became more a cultural rather than a utilitarian subject. In this regard, the teaching of every problem in mathematics should begin with a real-life problem. Relating school subjects to real life has a historical foundation. Tyler (1993) pointed out that a tremendous increase in the body of knowledge evolved with the advent of science and industrialization. Prior to that time school subjects were mostly academic, and there was little problem in selecting knowledge from the cultural heritage. When knowledge was needed for utilitarian purposes, however, content had to be selected from contemporary life. Knowledge in the real world should be utilitarian. Practicality, in both college and high school subjects, is essential to the efficient transfer and creation of knowledge.

The process of making knowledge relevant to life continued during the first and second World Wars when there was need for large numbers of skilled workers. Today, with the technological revolution and the constant application of knowledge to solve life's problems, it is more important than ever to make school subjects relevant to life.

Generally teachers teach mathematics as a cultural subject, emphasizing abstract concepts and symbols, with the result that students develop mathematics anxiety. "Many students view the current mathematics curriculum in grades [five through nine] as irrelevant, dull, and routine. Instruction has emphasized computational facility at the expense of the broad integrated view of mathematics that has reflected neither the vitality of the subject now the characteristics of students" (NCTM, 1995, p. 6).

Most students do not make the connection between mathematics learned in school and the application of that knowledge outside the school unless they learned the mathematics in a real-life context. Colangelo (1996) suggested that students need to be provided with the experiences that lead them to make these connections and apply school mathematics to real life.

Tyler (1993) suggested that students are more likely to be motivated when they see that what they learned in school relates to real life. In discussing the situation in which learning takes place, Tyler (1993) stated that this is more likely to perceive the similarity between life's situations when lessons and real-life intersect. He suggested that students will be involved in seeking illustrations in life outside school for application of things learned in school. Critics have pointed out that contemporary life is constantly changing; hence teaching children to solve problems today will not guarantee that they will be able to use the knowledge in the future, as the problems may be different. This will be the case if students are taught what to think, that is, specific content. If students are taught problem-solving strategies, and as Dewey suggests, "how to think" rather than "what to think," however, they will be able to transfer these skills to a variety of situations.

There is movement to change the teaching of mathematics from a content-oriented subject to one with a problem-directed empirical orientation, while not neglecting the abstractness of the subject. Theorists suggest that every concept in mathematics should begin with a problem, concrete representation, realistic situation, or other motivating stimulus. This concrete stimulus should be suitably analyzed so as to gain an understanding of its parameters and the mathematics needed to solve it. Data should be gathered and the teaching-learning process should end with generalization. Thus the lesson should lead from a realistic situation or problem to a real-life model and then to a mathematical abstraction.

In the current global society, there is greater need for personnel with skills in mathematics. Life is based on problems. If a problem is defined as an obstacle in the attainment of a goal, then every decision in life involves a problem. Learning theorists such as Gestalts suggest that problem-solving is the basis of learning. Every mathematical concept is based on a problem situation. The NCTM suggests that problems establish the need for new ideas that should serve as the context for mathematics in grades five through nine.

Fundamental mathematics skills are needed for basic survival in the global economy. Some practitioners believe that the use of concrete materials such as manipulatives and other materials should be reserved for slower students. This could not be further from the truth. Psychologists such as Piaget laud the use of concrete materials such as ice cream sticks and chairs for young children.

At a higher-level, however, concrete representations are useful for physical referents, for example, problems in construction. Algebra is usually thought to be a subject consisting of Xs and Ys and symbols like squares that have no meaning. Indeed Xs represent objects and squares represent physical space as in the Macy's example above in which a square represented the space for a display window. This was extended to a rectangular space that was described by a quadratic equation. Squares and triangles also may represent parts of a construction such as a bridge. In geometry every shape represents some object, and in calculus derivatives represent a fraction of a surface.

In the current global society, acquisition of mathematical skills is more critical than ever, both for investment purposes and for making everyday decisions. As the economy adapts to the information age, increasingly the workplace demands workers with mathematical skills in every sector. Hotel clerks, secretaries, auto mechanics, travel agents, and computer analysts need to know mathematics as much as investment bankers. Employers claim that most jobs require analytical skills such as the use of graphical, financial, and statistical data. Indeed quantitative data is presented in the newspaper, financial reports, and in every sector of the economy. Because schools have the role of preparing students for the economy, then the curriculum should involve mathematics that prepares students accordingly.

Mathematics is also part of everyday life situations. In business, decisions are made on activities such as calculating interest on a loan for a car or mortgage and interest on investments. Decisions on building houses, investments, traveling, shopping, and getting a job are all made on the basis of financial information. Students may use these everyday experiences in the solution of problems. Mathematical concepts in fractions, graphs, estimation, decimals, and percentages are used every single day. Situations from pupils' lives usually make good examples to serve as the basis of mathematics problems, for example, how to locate your house on a map, the way to a friend's house, or the way to school. There are also examples in business, such as the cost of a bicycle, or tennis shoe, or how much to save for a trip. Students can also play popular games such bingo or calculate the probability of winning the lottery. These concrete representations of everyday life situations and problems that should initiate every mathematics principle rather than the other way around.

The idea of using real-life objects and situations in teaching was exemplified by James Stigler and Harold Stevenson (1991), who did a comparative study of 120 classrooms in the United States and three Asian countries. The authors commented that "An American teacher announced that today's lesson concerns fractions. Fractions are defined, and she names the numerator and denominator. She asks, 'What do we call this? And this?'"

After assuring herself that the children understand the meaning of the terms, she spends the rest of the lesson teaching them to apply the rules for forming fractions. The authors commented that the Asian teacher does the reverse.

First the focus is on interpreting and relating a real-world problem to the quantification that is necessary for a mathematical solution. They gave a similar example from a Japanese classroom.

> The teacher began [by] posing the question of how many liters of juice . . . were contained in a large beaker. Children made several guesses and the teacher suggested that they pour the water into some one-liter beakers and see. Horizontal lines on each of the beakers divided it into thirds. The juice filled one beaker and part of the second. The teacher pointed out that the water came up to the first line on the second beaker, only one-third was full. The procedure was repeated to illustrate one-half. After stating that there was one and one-out-of-three liters in the first big beaker and one-out-of-two liters in the second, the teacher wrote the fractions on the board. He continued using fifths, etc. Near the end of the period the teacher wrote the word *fraction* on the board, and for the first time mentioned the words numerator and denominator. He ended the lesson by summarizing how fractions can be used to represent parts of a whole.

Thus the concept emerged from a real-life meaningful experience rather than from an abstract concept. The teacher's questions guided the children's thinking. From a real-life representation, abstract concepts were deduced, and the children understood the concept more easily. Asian teachers engage students in discussion of mathematics, using verbal information in the cognitive structure. Questions are for the purpose of divergent thinking rather than for a convergent answer.

According to Willoughby (1990),

> By deriving mathematics from the *learner's reality* and by constantly applying that mathematics back to real situations in which the learner is interested, we can help students to understand mathematics better and to see it as a useful, powerful, and even beautiful tool that helps to solve their problems and helps them understand the world around them better.

Experiential Background

The study of the brain has demonstrated that knowledge is anchored in the brain in the form of memories. These are recalled whenever there is contact with a new experience and associations are made. Thus past learning is a kind of anchor for new learning. It is like a rod on which a curtain is to be hung. Without relevant experience, learning of new materials will be arbitrary and difficult both to understand and to remember. Without the relevant experiences in the cognitive structure, a teacher has to provide the background information.

Every lesson has to begin with evoking knowledge from the cognitive structure of the students' past experience. Thus a useful beginning activity might be checking of previous knowledge, administering a diagnostic test to determine whether the students possess the relevant knowledge, or just questioning the students. Alternately the teacher might use a motivational activity that provides the relevant knowledge as the challenge or drive for beginning new learning. Experiential background may be the base from which the teacher selects problems for learning. It may be related to social class or culture of the students, in which case it becomes a powerful context to which the students may relate. Thus mathematics problems may be based on events or behaviors, or activities that occur in the sociocultural context.

Motivation

The term *motivation* is the drive or energy that makes one work. Psychologists have defined motivation as being based on a drive or need. It is drive when it is an extrinsic observable behavior determined by external circumstances and stimuli, and based on it one may make inferences. Motivation is need when it is intrinsic or determined by processes, such as curiosity, within the individual. In any event motivation is goal-directed and explains the reason that an individual performs certain actions. Several theories of motivation have been postulated, both extrinsic and intrinsic.

Theories of Extrinsic Motivation

These include cultural pattern, social field, and behavior. *Cultural Pattern* is proposed by anthropologists such as Margaret Mead who proposed that upbringing affects behavior and suggested that children's motivation is based on upbringing or cultural pattern. Kurt Lewin, the founder of *Social Field Theory* suggested that motivation derives from a totality of a field of forces, such as demands, attitudes, and groups, that surround the individual. These environmental factors create experiences that affect the behavior of students. Thus a teacher should ensure that the student has the relevant cultural and social background to lean a new concept.

Another theory of extrinsic motivation is based on Skinner's *Behaviorism*, which emphasizes the importance of reinforcement in learning. Behaviors are modified and reinforced when the correct responses are reinforced by praise, grades, and other symbols of success. Praise may be manifested by several means. The teacher may praise specific accomplishments by the students, draw the attention of other students to an accomplishment, use students' accomplishment in the teaching of another lesson such as a research project, and allowing students to make presentations about their accomplishments. Such recognition is considered reward and will motivate the students to greater success.

Extrinsic motivation is the one that is most commonly used in schools. It is a form of conditioning in that external means like grades, praise, and rewards are used to encourage students to work. Also motivating are use of information that is of interest to students and that is relevant to their social and cultural backgrounds. Whenever a teacher introduces a lesson based on something the students know, their interest is immediately evoked. The teacher's methodology also motivates students. When the teacher engages the students in practical activities students are ecstatic because they can create meaning from the task at hand. Activities that involve the students in cooperative and competitive activities also motivate them.

Theories of Intrinsic Motivation

These theories arise from inner tensions or needs of the individual. Two useful theories are the *Drive for Competence* proffered by R. W. White (1959) and *Hierarchy of Needs* by A. Maslow (1987). White proposed that students are motivated by the need to establish mastery over the environment. Satisfaction comes from attaining that mastery or control over tasks, such as by completing assignments correctly. Any new problem that is introduced into the learning situation causes disequilibrium, tension, and discomfort in the cognitive structure. Solving the problem confers competence and so the student returns to equilibrium and competence.

A. H. Maslow (1964) proposed a motivational system based on a *Hierarchy of Needs* that ranges through levels of physical and psychological needs. He distinguishes between deficiency or lower-level needs and growth or upper-level needs. Deficiency needs are related to physiological needs; safety; belonging, love, and affection; and esteem. Growth needs are self actualization, desire to know and understand, an aesthetic needs.

In implementation, the lower-level needs must be satisfied before the upper-level needs may be applied. Thus the teacher should observe the students, develop a profile of the students' social and psychological characteristics, and apply them in the teaching-learning process. The teacher should ensure that the physical conditions are satisfactory and that students get an opportunity to fulfill their social needs within the class situation. Only then can the teacher embark on a course of teaching students. Students are unmotivated when a subject is boring, when the methodology is mainly recall of facts, and when there is no activity. There are also problems of unclear directions, introduction of knowledge not relevant to the aptitude or sociocultural backgrounds, when there is conflict in the classroom, or when children do not see the need for achievement. The theory is that motivation works to arouse, direct, and sustain behavior. Hence the teachers' task in the teaching-learning process is to arouse and maintain motivation throughout the lesson. Motivation is affected by the readiness of the student to perform the task; hence the teacher must look to past experiences of the students. It is like a hanging a curtain. Attachment cannot be done without the presence of rods. The teacher must secure attention and create interest before beginning the lesson.

Motivation must also be maintained by allowing maximum participation throughout the lesson, by giving rewards, and by encouraging competition and cooperation. The teacher needs to inform students when they have succeeded and to give praise and approval. Also the teacher needs to check for understanding to ensure that the students have mastered the new concept. In other words, the concept must have been attained, and if there are problems with understanding, then the teacher should reteach the lesson. A positive learning environment is the surest way to maintain a vibrant class with students who are motivated to learn and succeed.

Attention and Interest

Interest is related to the aim of education and of the lesson. It is an intermediating position that "establishes the relationship between the child and the curriculum" (Brubacher, 1964, p. 238). This a kind of focus has garnered profound attention. Interest usually begins with a problem or a question that appeals to the learner and causes him to attend and begin some type of activity.

Interest becomes purposive in that it invites and directs the student toward some goal. By itself, instruction can be dull; therefore, the instructor has to resort to inducements to entice the students to learn. In this regard artificial inducement such as praise and grades may be used.

Interest may become a link between the learner and a dull curriculum. This is best done by intruding activities in which the learner is interested and using materials taken from real-life situations that surround the learner. Incentives from the home and community will make the activity meaningful. Even with real activities, the child will only be interested in those activities that have value to the child. I once supervised a student-teacher who conducted a class during which the children were very listless until the student teacher said, "ZZ Top." The students came alive and began to respond positively. I did not know what ZZ Top was at the time; later I learned it was a band. Interest made the students come alive. On another occasion, the teacher had to teach the students how to calculate the slope of a line. He used an example record sales by singers Ja Rule, J Zee, and the The Spice Girls. The students were much energized and quickly learned the types of slopes that corresponded to their record sales: positive in the case of the former artists and negative in the case of The Spice Girls.

Teachers would do well to take into account what motivational activities are meaningful to the learner. For example, video games, Pokemon™, baseball cards, and the like can all be used meaningfully in the teaching of mathematics.

Activities

Before engaging in a teaching event, the first thing a teacher should consider is, "What are the students going to do?" This is student-centered, activity-directed education. If the teacher's focus is on, "What is the teacher is going to do?", then it is a teacher-centered, authoritarian classroom, and the teaching-learning activities will be arbitrarily directed by the teacher.

The preceding paragraphs have described competing philosophies of teaching and various philosophies of teaching. The model of teaching chosen should be one in which students are actively participating and are actively using their minds to create knowledge. An experienced teacher once gave me advice, "Keep the children active, or else they will keep you busy."

Experience throughout my career has taught me that this advice is a maxim that every teacher should heed. Thus the most critical question that the teacher needs to ask is, "What are the students going to do?" Teachers should select activities according to that answer; there should be action. Some principles require consideration in making decisions about activities. The students must be engaged in some meaningful activity in which they are constructing knowledge:

- The activity must be based on some real-life experience.
- Emphasis should be on conceptual understanding.
- Problem-solving situations and strategies should be emphasized.
- Proofs should be proffered.
- There should be explorations and discovery of concepts.
- Hypothesis and algorithms should be followed.
- Resources must be used, that is, students must be using their hands in some way, such as manipulatives, analyzing and organizing data, making charts, using the computer or other technology, creating models, making or using concrete representations, or discovering relationships.
- The students should work cooperatively or individually.
- There must be higher-level questioning.
- Students must be involved in making generalizations rather than being given information arbitrarily.
- Students' activities must be based on some real-life experience.
- Testing should be by methods of authentic assessment and performance evaluation in addition to traditional methods.
- Every lesson should begin with a problem and end with a generalization rather than the other way around.
- Every lesson should begin with a real-life situation and end with application rather than begin with the rules and then end with application.

The following elements of the teaching-learning process should de-emphasized:

- Beginning a lesson with a definition and then using examples to prove a point.
- Memorization of principles without understanding.
- Teaching by telling.
- Using only convergent questions in testing.

The preceding principles should be considered in selecting activities for a class. The teacher should further consider in what pattern of interaction (i.e., as a group or by individual students) the activities will be implemented. Each instructional activity should be related to a pattern of interaction.

ASSIGNMENT

1. Examine the state's learning standards for mathematics and write a list of topics that may be used for a semester.

2. Explain with examples how you would use real-life examples in teaching mathematics.

3. Making references to real-life examples, plan how you would motivate a seventh grade class.

4. Select a topic in mathematics and describe the level of performance you would expect students to accomplish.

5. Tell with explanations what resources you would use to teach that topic.

6. Write what activities you would use to teach the topic you chose in question four.

Figure 9-1. The Bow of a Ship

Chapter 10

The Instructional Unit

There was a time, not so long ago, when the purpose of teaching mathematics was to cover the content in certain areas: algebra, geometry, and trigonometry. Each subject was allocated blocks of time. The objective of the courses was the control of the subject matter rather than the development and application of mathematical skills. Today, the objectives are different. The theory is that mathematics cannot be taught in a vacuum; therefore, teachers must organize content and learning experiences in such a way that skills and knowledge are the end product.

The Unit Plan

Learning experiences, including content and activities, are organized into units. The unit concept is based on the principle that subject matter should be regarded as an instrument rather an as an end in itself. This concept reflects life. When problems occur in life outside the classroom, solutions are discovered using many kinds of knowledge. Faced with a problem, the individual uses algebra, arithmetic, geometry, and trigonometry to solve it. In a sense human lives consist of a series of decisions and situations involving instability or disequilibrium.

Thus faced with a problem, the subdivisions of subject matter become indistinguishable. The emphasis is on the obstacles and the use of knowledge as in instrument in their solution. A unit should be planned around some theme or problem that requires different kinds of mathematics. The unit provides the teacher with the opportunity to motivate the students, and toward this end the teacher should make the problems a reflection of real life. First known facts are needed to solve the problem, and the outcome is new knowledge. Knowledge is both an instrument and an output of the teaching learning process (see Table 10-1).

Table 10-1. Structure for a Unit Plan Design for Instruction and Evaluation

The State's Learning Standards
Theme
Learner Characteristics
Objectives
Preassessment
Scope of content
Motivating Activity
Materials and Equipment
Teaching-Learning Activities
Formative and Summative Evaluation
Culminating Activity

A Design for Instruction and Evaluation

1. New York State Standards for Mathematics, Science, and Technology.

The teacher needs to select the relevant objective from the scope and seek content for a text to ensure that students are taught the current knowledge content. The standards are indicated above.

2. The Theme

The general theme expresses the central organizing principle of the unit. Literally, unit means *one* and the purpose of creating a unit is to have one idea that unifies the content so that all concepts are based on the theme and reinforce each other. In terms of learning theory, ideas have to be related so that each new topic evokes the preceding one. Further, mathematics by its very nature is organized as a set of sequential principles that build on each other. Not one may be missed. A theme is derived from real life provides opportunities for students to relate and make mathematics meaningful. For example, a unit on the supermarket would provide the context for rate, proportions, weights, measurements, and money.

3. Characteristics of the Learner

The teacher is not only teaching a subject matter content, but also the students. The teacher needs to take into consideration the background of the learner such as:

a. *Academic background* including grade point average, and previous learning, and
b. *Learning styles* including whether students are independent or dependent learners who need structure. They may require direct guidance by teachers in how to perform a given assignment or direct instruction. Teachers should assess whether the student is effective at working cooperatively. Knowledge of the student's modal capacity (whether students are visual or verbal learners) will help the teacher select and prepare the mode of delivery. Knowledge of the needs and interests of the students will assist in preparing assignments that will motivate students.

4. The Instructional Objectives

Objectives should be written at one of the various levels of the taxonomy described by Bloom (1999).

This is consistent with the general principle of teaching that states that knowledge should be presented to students in progression from the simple to the complex. Bloom's taxonomy is a more sophisticated principle that ranges from recall of facts to comprehension, application, analysis, synthesis, and evaluation. Objectives written in accordance with these principles ensure that the unit will begin with the easiest lessons and proceed to the more difficult.

The objectives should be written a much as possible in conceptual form for different levels of the taxonomy. The organization of objectives ensures that easier lessons are taught first and that they will become progressively more difficult. Moreover, the writing of the objectives in developmental form sets the stage for linking the lessons with evaluation by testing. Another method of sequencing objectives is based on the "varieties of learning" model developed by Gagne (1985). Gagne proposed that there are levels of cognitive functioning and types of knowledge suitable for each aspect of mental behavior. Learning is arranged in a hierarchy as follows:

a. Factual learning of basic terminologies and facts;
b. Conceptual learning or concrete concepts that involve identifying spatial relations and classifying objects by characteristics;
c. Principle learning or defined concepts that require learning the relationships among concepts; and
d. Problem solving involving applying the principles just learned to new situations, making generalizations, inferences, predictions, and evaluation.

The types of learning are progressively more difficult, and each new type requires mastery of the previous learning. Objectives based on learning sequencing such as Bloom's taxonomy ensure that the content is presented in progressive order of difficulty.

5. Scope of Content

The scope of content should be:

a. Written in sequential form in order of difficulty so that each principle is a framework for the preceding one. The sequence should be logical.
b. Beginning lessons should use concrete examples and gradually become more abstract.
c. Early lessons should be in the acquisition phase of learning and the later ones should apply the principles.

6. Preassessment

Mathematics is based on a hierarchy of knowledge. Any missed principle creates a block in learning. The teacher needs to know whether students have the prerequisites for learning the new knowledge. Diagnostic tests based on the competencies that will be required in the new content therefore should be used to inform the teacher of the students' accomplishments within the hierarchy. Questioning is one option for acquiring this information, but a short test is the most effective instrument for such screening.

7. Materials and Equipment

This is a common-sense preparation. Teachers should know ahead of time rather than finding out as they begin a lesson that the school is not in possession of a video tape recorder or that the overhead projector is not functioning. Effective planning includes acquiring any materials required or arranging for alternatives.

8. Motivational Activity

A unit should take about two weeks of teaching time. A strong motivational activity at the beginning provides motive force to the unit. The motivational activity should portray a comprehensive understanding of the theme as a whole. Motivational activities also give direction to the unit. Activity options include field trips, bulletin boards, guest lectures, and research in the library or on the Internet. A visit to the planetarium may precede a unit on spatial relations and angles. The motivating activity should be evident in the classroom for the duration of the period.

9. Teaching-Learning Activities

Teaching-learning activities should reflect their description: they should be activities and they should have the following characteristics:

a. They should be based on real-life experiences of the students and of society. The traditional method of teaching requires the teacher to teach the content first followed by application to problems based on the textbook. A better method of teaching is to create the real-life experience first and from it to deduce the mathematical principle.
b. They should be real activities. They should not consist of a teacher "talking and using chalk" while the students sit and absorb. Instead, the mind should be actively engaged in creating ideas and giving meaning to the knowledge. Not

only should the students' minds be active but also their hands. Thus manipulatives, real-life representations, and situations should all be the basis for understanding a principle. For example, for a lesson on area, a teacher may have models of circles, spheres, and cones.

c. There should be some kind of resource such videos, computers, projectors, manipulatives, and programmed instruction materials. There should also be materials such as samples of telephone bills, newspapers, books, pictures, and photographs. Materials such as balance scales, currency, and coupons and teacher-made materials such as paper for making representations and worksheets can add interest and a connection to real life. The teacher should ensure that there are sufficient resources and materials for everyone. Whether the classroom is organized by whole groups, small groups, or individual students, each student should have access to available materials.

d. The activities should reflect what workers do in life. An investor calculates interest rates, an architect constructs buildings, and a housewife manages a household by shopping, planning, and organizing. This is consistent with the principle that real-life activities should be the norm for classroom activities.

10. Evaluation

Evaluation measures the effectiveness of outcomes of the unit. Effectiveness is determined by the performance of students on tests or other measures of evaluation such as authentic assessment. The objectives guide the selection of an evaluation type. Evaluation should conform to the hierarchy of instructional objectives, that is, they should include recall of facts through problem-solving. In this way evaluation is linked to instruction. Evaluation result should reveal what corrective actions should be used in the case that objectives have not been obtained. Both the students and the teacher should be evaluated, the students by tests and the teacher by reflective practice

Evaluation should be both process evaluation and product process evaluation, which is generally called formative or ongoing evaluation. A teacher does not need to know at the end of the lesson that the students did not learn but should constantly monitor the students' progress by tests, quizzes, and other means. Product evaluation occurs at the end of the unit and may be a test or product such as a projector display.

Authentic assessment provides an activity based on a suitable real-life experience for the student to enact. Performance assessment should be applied when the student has performed some act.

Alternative assessment (authentic assessment or performance assessment) include, for example, portfolios or a project. Portfolio is an ongoing means of assessing students in which the best works are revised, so that only the best works are displayed. Portfolios represent longitudinal efforts by students.

11. Reflection

Reflection is the final stage of learning in which teachers look back and evaluates themselves. They examine the suitability of the theme and determine whether the objectives were attained. If the objective was not attained, strategy revisions should be applied before beginning the next unit. Reflection should take place at the end of every lesson and review should be a continuous process.

12. The Culminating Activity

The culminating activity should be a dramatic representation of what has been accomplished in the unit. It should be a product such as a display, dramatic presentation, mathematics fair, presentation, or other interesting representation of the unit. It should be comprehensive and be presented by the students themselves to allow them to share their new knowledge and demonstrate what they have learned.

Sample Plan

The following provides a framework for the construction of a unit plan.

State Standards:
The state standards represent the objectives of the lesson. There are general and specific standards. The teacher should first select the standards and then seek the content to be taught from a text book.

Theme: The city plans to rebuild on the World Trade Center Building Site.

Learner Characteristics:

1. An eighth grade of mixed academic ability; twenty-five students with IQs of 95–130.
2. Generally middle class multicultural background.
3. Average reading level.
4. Highly motivated college-bound group of students.
5. Students showed an interest in geography and architecture.

Conceptual Objectives

1. To ensure that students know how to use rates to relate to quantities measured in different units.
2. To enable students to see the coordinate system as a system that relates two or more variables.
3. To help students to understand that a graph is a visual representation of the relationship between two variables.
4. To help students apply known concepts such as area and perimeter.
5. To help students grasp the idea that a slope represents a rate of change or the line that represents the number of units the vertical line rises or falls for each unit of change of the horizontal unit.
6. To help students apply the slope and the coordinate system in the intercept form of the equation in the solution of practical problems.
7. To help students to draw to scale.
8. To use technology to perform calculations. Excel spreadsheet will be used for calculation of the telephone bills.

Scope of Content

1. Rates.
2. The Cartesian Coordinate System.
3. Creating graphs.
4. Slope as a rate of change.
5. The slope-intercept form of an equation.
6. Applying the slope to real-life problems such as calculation of telephone bills.

Preassessment

Students have experience with the number line, order of operations, and solution of simple equations.

Students will be given a diagnostic test for the following competencies:

1. Performance of order of operations.
2. Writing of simple equations.
3. Knowledge of the number line indicating positive and negative numbers and zero.

Materials and Equipment

1. Copies of telephone and electricity bills.
2. Map of the world indicating latitude and longitude.
3. Bills from the supermarket, gas sales; salary coupon indicating rates.
4. Photograph and biography of Descartes, Bill Gates.

Motivating Activity

Students will take a field trip to World Trade Center Site in New York City. They will gather drawings of the area visited.

Culminating Activity

Creation of map of the site to be used for construction indicating the site of buildings for a World Trade Center Replacement Building and calculating the slope of the gradient. Students will work in four groups and display four maps. Includes constructing buildings to scale, calculating distance and rates, calculating the costs of telephone bills and transportation costs by car and train for the trip, writing the budget for the project, and display of products.

Evaluation

Process: Students will be given a quiz or a worksheet at the end of each lesson.

Product: Project

1. A short biography of Rene Descartes.
2. Maps of the construction site produced by groups of students.
3. Computer-generated telephone bills for families.
4. Transportation costs of trip.
5. The budget for the project.

All products will be displayed.

Reflection

Students and teacher will make a summary of what has been learned and make a chart. Teacher will cite gaps in knowledge and rectify the shortcomings. The teaching-learning process is circular. It begins with objectives, implementation, and feedback. Reflections are the feedback that improves the work and makes the product into a work of art. Calculating Slopes in Real-life Situations and Costing Projects are examples of *teaching-learning activities* in a unit plan as shown in Table 10-2.

Table 10-2. Unit Plan Teaching-learning Activities

#	TOPIC	MAIN CONCEPTS	ACTIVITIES	EVALUATION
1.	Rates	Rate as an average	Plan cost of trip consider speed and distance	Make a plan of a activities for a field project
2.	Applying Rrates	Solve problems using rates	Shopping list for equipment	Accuracy of computations
3.	Coor-dinate System	Main terms: origin, axes quadrant	Create a grid for locating places in field. Simulate w/tape strips on classroom floor. Record table of values	Make a map of the schoolyard w/location, e.g., auditorium
4.	Graphing ordered pairs	Positive Negative Zero slopes	Plotting chart of sale of equipment e.g., phones, CDs. Compute tables	Record sales Comparing Interpreting Slopes of values
5.	Slope of line	Slope as a rate of change	Climbing up a ski Calculating change Calculating slope $S = \frac{\text{Change Y}}{\text{Change X}}$	Recording time, distance, using Excel
6.	Slope/ intercept	Formula	Use real-life examples, e.g., Cost of phone calls. Deduce formula $Y = A + BX$ Translate to $Y = MX + B$	Calculating phone fills for families and phone co., use Excel
7.	Linear Equations	Formula	Applying linear Equations to Problems	
8.	Unit Test			

ASSIGNMENT

1. Write a set of instructional objectives for a semester.

2. Choose a theme and write a unit plan for a two-week period.

Chapter 11

Preparing and Implementing the Lesson Plan

The previous chapter discussed the principles of planning for instruction and outlined a unit plan. This chapter discusses specific mathematics pedagogy. Mathematics methodologies are based on the learning theories and approaches to teaching that have been discussed in the preceding chapters. The aim of teaching every lesson is that students learn something new (Blair, 1998) and that this new knowledge builds on the knowledge existing in the students' cognitive structures. Just as an architect first creates a plan before he executes it, so must a teacher create a lesson plan. Teachers do not generally like to make plans, but experience also counts for planning. A lesson that is not planned cannot be executed effectively. An individual lesson is the basic element of teaching (Blair, 1998). Each lesson should be based on the following components:

1. State standards;

2. Instructional Objectives: cognitive and performance objectives;

3. Materials and Resources;

4. Instructional Objectives such as discovery, problem-solving, and cooperative learning;

5. Procedures

A. *Background and previous knowledge*: Knowledge of the students' academic, cultural, experiential, and socioeconomic backgrounds as well as previous knowledge are basic to ensuring that students are ready to learn the subject.

B. *Motivation*: Usually a problem or real-life situation.

C. *Developmental Activities*:

Subject Matter Content: Should be arranged in logical sequence sothat one component is taught at a time, ranging from facts to generalization of concept.

Activities: Facts should be taught through activities by students' use of resources, and questioning by teacher rather than by telling.

Generalization: Summary of key points or formula should be deduced from students and recorded on chalkboard.

Practice: Structured, guided, and independent practice follows.

6. Summary/Evaluation

Assessment may be through quiz. game, group activity, worksheet, or creation of some product.

7. Teachers' reflections and follow-up.

The Lesson

The focus of this section is to demonstrate major methods of planning and delivering instruction with appropriate examples. It discusses the standards of the National Council of Teachers of Mathematics

(NCTM); states' learning standards; writing objectives; and methods of organizing and delivering content, including the logical and psychological orders of a lesson, activities, and questioning.

Learning Standards

Standards for mathematics have been outlined by NCTM and based on these standards are state learning standards promulgated by New York State. New York State presented an integrated method of mathematics called Mathematics, Science, and Technology. Relevant standards are outlined below. NCTM has promulgated standards for grades seven through twelve. These are standards of content and standards of process.

The mathematics department plans the year's curriculum based on the learning standards. The individual teacher uses this plan as the basis of lesson units and from this individual lesson plans. A typical lesson plan has the following components (2000 Curriculum Standards for Grades K–12).

Content Standards

1. Number and Operations
2. Patterns, Functions, Algebra
3. Geometry and Spatial Sense
4. Measurement
5. Data Analysis, Statistics, and Probability

Process Standards

6. Problem-solving
7. Reasoning and Proof
8. Communication
9. Connections
10. Representations

Writing Objectives

The objective is the goal of every lesson. It allows the teacher to measure the degree to which students have mastered the content. Thus objects are a measuring or assessment standard that determine when the students have learned the material. Thus they guide the teacher toward a goal.

Objectives for a lesson are derived from the New York State's learning standards, which outlines the objectives in conceptual form. The teacher selects the relevant standard, builds learning units from them, and then creates lessons from these units. Decisions about objectives are the professional responsibility of the teacher and reflect the teacher's training in pedagogy. The objective should be described in conceptual and performance terms.

Conceptual Objectives

Conceptual objectives describe the main concepts to be taught. Remember that these objectives are in the cognitive domain and should be written at one of the levels of cognitive functioning as described by Bloom. These are shown in Table 11-1.

Table 11-1. Cognitive Objectives Based on Bloom's Taxonomy

Level	Objective	Description
1	Knowledge	Remembering of facts.
2	Comprehension	Understanding the meaning of at least two relations.
3	Application	Making generalizations.
4	Analysis	Breaking down material into component parts and examining relationships such as comparison.
5	Synthesis	Combining parts into a new meaningful whole.
6	Evaluation	Making judgments about relations relations based on given criteria.

Mathematics has a well defined body of content that must be taught in segments. Conceptual objectives describe the knowledge that must be taught in a particular lesson. Statements of the objectives begin with words that introduce the knowledge orientation of the lesson, such as understand, appreciate, grasp, and know. These words denote understanding of the concept. For example, in a lesson teaching decimals, a conceptual objective could be stated as

- Students will understand
- The group will know
- May and Susan will appreciate
- Each student will grasp

In sum, the objective would read, "Students will understand that decimals a fractional part of a whole with value system based on tenths."

Performance Objectives

As the word implies, performance objectives or instructional objectives describe the particular action that the student must be able to do at the end of the lesson. The performance objective should be comprehensive and specific. Group-Activity-Situation-Performance (GASP) is an acronym for describing performance objectives:

G is the individual or the small or large group to which the lesson is directed (i.e., Louisa, Group A, or the class),

A is the activity to be done by students,

S is the situation or medium in which the action is to be performed, and

P is the performance standard, such as accurately or one hundred percent.

The objective must begin with a verb, which denotes action such as write, present, differentiate, solve, construct, compare, contrast, and list. For example, if the lesson is on decimals, the objectives could be stated thus:

- Find the unknown variable.
- Apply the formula to the problem.
- Identify the place decimal point.
- Reconstruct the fractional value.
- Formulate a real-life problem.

The third aspect of the objective is to identify the situation or circumstances under which the action will occur and the type of assessment procedure. Specify skills to be mastered and methods of identifying them, such as a quiz, test, or worksheet. In considering product evaluation, identify the product such as test or a project. For example, on the teaching of decimals the medium may be a worksheet, a class presentation, or on a quiz.

In total the performance objective may be described thus: "At the end of the class students will be able to calculate correctly on a worksheet the dollar value of three bills with five items bought at the supermarket."

Materials and Resources

Every lesson should utilize some type of material. Resources help to demonstrate and explicate the concept. Materials should be based on real-life examples such as scissors, scales, money, coupons, geometrical shapes, and models.

Organization and Delivery of the Content

Whether the approach is authoritarian or child-centered, the lesson content has to be organized in some way. In logical organization the order is determined by the nature of the subject discipline. This means that the matter is organized deductively in a hierarchical sequence. The lesson is organized psychologically when the order is determined by the individuality of the students, that is, by psychological maturity or interests (Brubacher, 1969).

Content

The logical order of the lesson assumes subject-centered curriculum. The role of the teacher is to incorporate this new knowledge into the minds of the students. This assumes an authoritarian model; both the teacher and the student must obey the objective order of the content. Teaching is based on the point of view of the person who already knows the content (Brubacher, 1964), that is, the teacher. Logic proceeds from the simple to the complex or follows the order of nature. For example, in teaching chemistry, the order begins with electrons and proceeds hierarchically to protons and neutrons, to an atom, and to a molecule. Critics of the authoritarian methodology contend that logical methodology fails to motivate the students and encourages memorization.

In the logical order model, the content is divided into sequences and each taught at a stage of the lesson. The process is based on deductive logic and is suitable for lecture or direct instruction methodology. It begins with definitions and continues with examples to explicate the definition.

The Psychological Order of the Lesson

The psychological order of the lesson is based on how children learn. Instead of starting with the content, organization begins with the interests and needs of the child (Brubacher, 1964). It takes into consideration their maturity, past experiences, and hence readiness to do the task. It may also begin with a problem that causes disequilibrium in the cognitive structure of the students. The organization of the content nevertheless has some logic. Usually the lesson to be learned by the student is already known by the teacher and generally the student's task is to reinvestigate and discover what is already known.

The Teaching-learning Process

Concept attainment is central to the teaching-learning process using the psychological order of the lesson. By definition, whether a concept is concrete or defined/abstract, it has criterial attributes. Teaching involves guiding students to follow the method of concept attainment. This involves learning one attribute at a time and then ending with the concept name or generalization. Delivery of the content is based on how the mind learns, by processing ideas through accessing previous knowledge, making associations, by analyzing relationships, organizing information in patterns, making generalizations, and rehearsing and reinforcing. By definition, every subject has a structure that governs or explicates how the main ideas are related. Leaning the subject involves discovering the structure of the subject matter. It includes methods of problem-solving and discovery.

Teaching Strategies

Two strategies should be used in every individual lesson. These are (1) activities and (2) questioning.

Activities as Teaching Strategy

The lesson is geared to teach the student something new. Therefore the first question the teacher needs to ask is, "What will the students do?" What is the activity? This should be actions in which the students are involved in using their minds, hands, and resources. The content of every lesson should be divided into components or facts. Each fact should be taught through an activity using resources. For example, if the teacher wishes to teach a concept that has three criterial attributes, the teacher will divide the lesson into three facts and three activities:

Fact 1
Activity

Fact 2
Activity

Fact 3
Activity

Generalization

Generalization applies to the concept that will be deduced by the students. This method of concept attainment and activity has been used by teachers in child-centered education since the time of Froebel and Pestalozzi in the nineteenth century. It is no less useful now.

Questioning as Teaching Strategy

Main purposes of questioning are twofold:

1. To test what students know, that is, knowledge existing in their cognitive structures that are the foundation on which the lesson is

to be built. Such knowledge is elicited by questions beginning with who, what, where, and when;

2. To stimulate thought, that is, to engage the mind in an active creative process, so that it can perform the functions of analyzing, discriminating, associating ideas, and making generalizations. Such questions begin with how and why.

Questioning may be classified by the level of functioning of the brain (see Table 11-2). The classifications are

1. Closed or Convergent questions request one answer. These are two choices, right or wrong. These elicit knowledge and not thought and are suitable for recall of facts or comprehension.
2. Open or Divergent questions stimulate thought. They invite the mind to think and create answers that convey new meanings. They are suitable for lessons that deal with discovery and problem-solving.
3. Cognitive or Affective stimulate mental functioning based on brain activity. Cognitive questions evoke knowledge that has been organized in the memory of the brain. The concept of cognitive functioning has been described by psychologists. Benjamin S. Bloom (1999) has been credited with the classification of cognitive objectives in his edited book titled Taxonomy of Educational Objectives. The categories range from recall of facts to evaluation. Affective questions deal with emotions or feelings such as attitudes and impressions. Cognitive functioning correlates with emotional behavior in every individual.

The aim of asking questions is to enable the individuals to use the cognitive structure, that is, the sum of knowledge organized in the memory, and to apply this knowledge to new problem-solving situations. The interaction of the amount of knowledge or native intelligence will guide the creation of new knowledge. The guide through this cognitive structure and the quality of the knowledge deduced will depend of the quality of the questioning. Generally questions will direct the functioning of the brain and will help the individual to associate new and existing knowledge, make generalizations, and solve problems. Thus questioning is an important skill in the teaching-learning process.

Table 11-2. Levels of Cognitive Functioning and Questioning

Level	Cognitive Process	Mental Functioning	Lead Word
Lower	Knowledge of	Facts Terms Definitions Concepts	Define Describe List, Name Repeat
Middle	Comprehension	Understanding of ideas	Classify Explain Interpret Summarize Translate
	Application	Apply new skill to the solution of a problem	Apply Construct Compare Simplify Solve
	Analysis	Breaking into component parts and showing relationships among parts	Compare Contrast Distinguish Identify Relate
Upper	Synthesis	Putting together parts broken by analysis and creating new whole	Change Create Explain Generalize Integrate Produce Summarize
	Evaluation	Making judgment about standards based on synthesis	Assess Conclude How well? How effective

Benjamin S. Bloom compiled a taxonomy of educational objectives based on the cognitive domain. The objectives are classified into a hierarchy based on the level of mental functioning and range from knowledge at the lowest to evaluation at the highest. In addition to basing tests on this hierarchy, classroom questioning may be based on educational objectives depending on the purpose for which the questions are asked. The levels of cognitive functioning and the lead words that elicit that function are listed below.

A hierarchy of mental functioning based on Bloom's taxonomy and corresponding lead words that may be used to evoke suitable responses are listed in the model.

Both deductive and inductive questions should be used in the teaching process. Questioning can make or ruin a lesson. It is the heart of the teaching process. Any teacher who wishes to teach well should learn to ask questions. The poignancy of questioning was demonstrated by Jesus Christ and by Plato. Jesus, founder of Christianity, taught the parable of the Good Samaritan. The rich young ruler asked Jesus, "Who is my neighbor?" Instead of declaiming the definition in the usual way, Jesus told him a story of the Good Samaritan, the man who went from Jerusalem to Jericho and fell among thieves. The passersby responded with degrees of indifference. It was the Good Samaritan, an outsider, who rendered aid to the man that fell among thieves. Jesus then asked the question, "Who then was neighbor to him that fell among thieves?" The listener was able to give the definition. The method used by Jesus was the inductive method, which followed the sequence:

- *Question*, or statement;
- *Analysis*, or the story; and
- *Generalization*, or the listener's definition at the end of the story.

The inductive method was used by Socrates (469–399 B.C.E.), who taught the slave boy how to find Pythagoras' Theorem. The steps are:

- *Socratic irony*, or statement;
- *Definition* of the statement identifying the criterial attributes of the concept;
- *Analysis* of the definition, involving examination of each criterial attribute, e.g., by comparison, contrast, examples, and nonexemplars, and analogy; and
- *Generalization*, such as rule, principle, or concept name.

The inductive method implies that the learner has some basic concepts within the cognitive structure. Where the specific knowledge is absent, the teacher may make analogies. The inductive technique is the best for teaching by making the mind active and stimulating mental functioning. Teachers may well learn from the examples of Jesus and Socrates how to use the method of questioning effectively.

Types of Questioning

For then purposes of the teacher the following types of questioning may be used.

Focus Questions draw attention to the lesson. In the teaching-learning situation, there are several stimuli in the environment. The mind can effectively process only one at a time. Focus questions draw attention to one stimulus so that the mind is able to concentrate. The teacher's responsibility is to phrase questions so that the mind can focus on one idea at a time. These questions are suitable for recall, for facts to check what the students' previously know, and to focus attention on the day's lesson.

Prompting Questions, as in drama, are a prompt, cue, or hint. Students may need such a cue to help them recall facts in the cognitive structure. Thus if a teacher is asking a question and the answer is not forthcoming, the instructor should use prompting questions to trigger memory. Prompting questions may be used as a follow-up to a previously asked question, to correct wrong responses, and check levels of learning.

Probing Question are designed to investigate some phenomenon in depth. The process is developmental in nature. Probing questions are used to clarify unclear ideas and to develop a train of thought. Knowledge is organized in a develop mental manner ranging from the simple to the complex. Probing questions seek to investigate and clarify ideas from their nascent beginnings, through stages of complexity, until the students attain insight. It is indeed the heart of teaching.

Questioning Skills

To attain the desired objective, the teacher must practice questioning skills.

Redirect is used when a student gives the right answer at the first try. The teacher tells the students, to keep the answer for a while after which he returns to the first student. In the meanwhile, the teacher asks the same questions of other students. This allows for maximum participation by students.

Wait-time is a pause. The teacher should ask the question and then pause to allow students time to think. It is important to allow students to complete their thoughts.

Reinforcement shoud be used whenever a student gives a right answer, the teacher should reinforce with praise. Reinforcement may also take place by asking several students to repeat the answer.

Questioning Guidelines

The teacher should:

Pose clear and unambiguous question. Vague question like, "What do you find in the ocean?" should be avoided.

Ask the question before designating a respondent. This allows all the students to think. If a respondent is designated before the question is asked, the rest of the class may not participate in thinking. Ensure that the questions match the objective of the lesson.

Match the question to the ability level of the student. It is useless to ask abstract questions of students who are operating at the concrete level.

Avoid rhetorical questions, which are statements ending with, "Isn't it?" Rhetorical questions do not require thinking.

Avoid postscript questions. An example is: The area of a circle is radius squared by pi. What is the area of a circle?

Avoid repeated questions. The teacher should not repeat questions, which may prompt the students to wait for the questions to be repeated starting to think.

Avoid multiple questions, that is, questions that require several answers.

Dealing with Answers

Dealing with answers is just as important as asking questions. The following are useful questioning skills. The teacher should

Indicate whether the answer is right or wrong. The teacher should show appreciation for even partial answers. In this case, it is possible to defer giving the answer to the question until other answers are put forth, but it is important to respond to each student who offers an answer.

Explain why an answer is wrong, perhaps by asking follow-up questions or asking another student to explain a right or wrong answer. The teacher should not answer the questions himself and should not habitually repeat the question.

Ensure that the question is heard by the students in the class.

Every teacher should practice the skills of good questioning. The teacher should use focus, prompting and probing questions. He should practice asking clear questions, ensuring that he or she asks the question before a respondent is designated, and deal with answers effectively.

How to Teach Using Questions

The following lesson demonstrates how questions may be used in teaching a lesson.

1. Understand and analyze the problem or situation
2. Divide the content into a sequence of facts to create a real-life model.
3. Base activity on the use of concrete resources.
4. Use questions to elicit the facts.

5. Translate the real-life model into a mathematical model.
6. Help students make a generalization.
7. Provide structured and guided practice.

Lesson Plan to Demonstrate Questioning Skills

1. *Topic*: Area
   Grade 7

2. *Objective*
   a. Conceptual: To help students understand that area is surface.
   b. Performance: To enable students to calculate the area of carpet given the dimensions.

3. *Instructional Strategy*: Inductive/concept development

4. *Materials and Equipment*: Plastic tiles or squares of paper

5. *Procedures*
   A. *Previous Knowledge*: Students know the concepts of length and width
   B. *Motivation*: Problem: John wants to place the carpet on his floor. He wants to know how much carpet to buy. What advice would you give to John?

   C. *Developmental Activities*
   Fact 1: To know that area means surface.
   Activity 1: Pose the problem using the classroom table as a model of the floor and ask the following questions:

   Q. How would John know how much carpet to buy for the room?
   A. Measure it.
   Q. How is carpet measured?
   A. By square feet
   Q. This is my room; where would I measure?
   A. The surface of the room indicated by the table (touch it).

Q. Inform students that another name for surface is area, or they might know the concept name. Identify several other surfaces, for example the floor or chair.

Fact 2: Area can be found by calculating the dimensions of the surface.

Activity 2: To create a real-life model of area. Students will be divided into four groups with relevant materials. Each group will perform an experiment using their desks as the room and the squares as tiles. Students place the tiles on the tables to cover them.

Q. Count how may squares have covered the desks.
A. 25 squares.
Q. If 25 square feet covers the desk, what is the area of the desk?
A. 25 square feet (emphasis that area is square).
Q. How did you arrive at this answer?
A. Count the length and width and multiply. (Reinforce by redirecting.)

Generalization: Area can be found by multiplying length by width.

Q. Now suppose that I had a large room. Would I be able to count squares?
A. No.
Q. Suggest a way that area could be found without counting squares.
A. Measure the length and width.
Q. How can the area may be found mathematically?
A. Measure the number of squares on the length and width and confirm that area

L = 5 and W = 5 Area = 5 * 5 = 25 sq. ft.
Length = X and Width = X Area = X^2

Reinforce by allowing students to repeat generalization and by writing the formula on the chalkboard.

6. *Application*. Refer to original problem.
Q. Now based on your experiment, what advice would you give to John?
A. Measure the length and width of his room to find the area.

7. *Practice.* Individual and group practice applied to rectangles.

   Problem 1:. How many square of carpet is required for the top of your desk? For your classroom?

   Problem 2: Given that the dimensions of John's room is 75 feet long and 50 feet wide. How many square feet of carpet would he buy?

   Problem 3: Draw a model of your own room and calculate how many square feet of carpet would be needed to carpet it.

8. *Summary/Evaluation*

   Of course questioning has to be spontaneous. The questions cannot be fixed beforehand in case the students do not give the expected answers. The important thing for the teacher to know what facts are being deduced and to ask the questions accordingly until the expected facts are *elicited* and *generalized.*

ASSIGNMENT

1. Discuss the purposes of questioning.

2. Name the types of question and lead words that could be used to accomplish the learning of specified content.

3. Work in groups. Choose a concept and write a series of questions to use to teach a concept.

4. Make a lesson plan and teach it using questioning.

Chapter 12

Models of Delivering Instruction in Mathematics

This chapter discusses specific models of teaching mathematics as a cognitive process that uses the inductive reasoning. These are

- Problem-solving/discovery,
- Real life problem situation,
- Concrete representation, and
- Concrete/discovery.

Problem-solving/Discovery

A problem is an obstacle in the way of attaining an objective. In this regard, the incidence of problems in life is perennial. In the past, problem-solving was used to mean a topic in mathematics, just like addition or geometry. It was taught to children and practiced like other topics, and assessment was done by giving students problems to solve (Kennedy & Tipps, 1994). This was based on the philosophy of mental discipline that has been previously discussed. Today the meaning of *problem* has a different connotation. Evolutionary science, which inspired the epistemology of pragmatism, conceived that the essence if live is solving problems. Problems exist in the environment, and man solves them daily in the process of adjusting to the environment. Problems stimulate thinking, and according to Dewey, thoughts are plans of action that arise as man encounters problems in the environment.

Problem-solving is therefore a *cognitive process* that emanates from the social environment. Life is defined by problems. Pragmatic theory teaches that the curriculum should be problem-centered and based on societal problems and that the purpose of education is to teach students how to think rather than what to think. Learning theory evokes ideas of problem-solving as a method. *Behaviorists* conceive that learning occurs when there is a change of mind because the environment changes and becomes problematic: problem-solving is a response to stimuli in the environment. When the stimulus occurs, the individual responds by reorganizing the environment (Brubacher, 1964). Although behaviorists do not suggest what happens within the brain, cognitive psychologists do.

Cognitive psychologists conceive of problem-solving as a process involving the functioning of the brain. Beihler and Snowman (1997) suggested that cognitive processes play a central role in problem-solving. Cognitive psychologists such as Jerome Bruner (1960) discussed the matter of problem-solving and discovery. In his book, *The Process of Education*, Bruner proposed that teachers should help students to understand the meaning of structure in learning the subject. Understanding structure means understanding the basic or fundamental ideas and how they relate to each other. Structure can be represented by a diagram, picture, verbal statement, or formula. An example is Einstein's formula for relativity. Bruner also stressed discovery in problem-solving.

Bruner (1960) questioned the step-by-step approach to teaching that is commonly used in schools. He suggested that the method leads to student dependence and that learning should be presented with a problem to which students seek solutions. A problem creates disequilibrium in the brain and challenges the students to employ a process that leads to a return to equilibrium. To describe the process, Bruner discussed the situation in which the students were asked to explain the purposes of a compass. One student explained that it is a tool. Other students quickly pointed out that there are other devices that could be useful for the same purpose, such as a camera tripod. The importance is that in presenting a problem, students need to see the whole situation so that they can discover relations.

Robert Gagne (1985) suggested that problem-solving is best accomplished with conditions within the learner and those in the learning situation.

Conditions Within the Learner

The learner should have within the cognitive structure certain facts that may be recalled. The facts or previous experiences are the anchoring ideas without which learning cannot occur. Without those anchoring ideas, learning would be like trying to hang curtains without a rod.

Within the cognitive structure there must be verbal information organized in the memory, usually in a schema. The problem situation provides a cue that links the existing knowledge to the schema. Thus the learner will be able to make connections and develop a strategy for the solution to the problem. Learners are well served if they within their cognitive structure the skills with which to conduct a search for information.

Conditions Within the Situation

External conditions must be amenable to problem-solving. For example, the teacher gives verbal instructions that act as stimuli and prompts the recall of facts. Verbal instruction guides the learning channels in the brain. Problem-solving involves discovery, which requires the learner to use higher-order thinking skills to construct a new rule. Recent research on instructional methods involving exposition and discovery has shown that exposition promotes for superior recall when a measure immediately follows an event. Transfer is the more important outcome of learning. Gagne (1985) suggested that problem-solving leads to

1. The acquisition of intellectual skills, concepts, and rules. Previously learned materials are retrieved from the memory and brought to bear on the new situation. New learning is the outcome.

2. Learning is guided by cognitive processes, therefore, in examining relationships. Problem-solving is discovery. The learner does this using verbal information.

3. The learner must have internally organized knowledge that is brought to bear on the learning processes. These are cognitive strategies.

All three capabilities—intellectual skills, verbal information, and cognitive strategies—are absolutely essential in the problem-solving. It is the teacher's responsibility to guide the students through this maze by (1) presenting materials that have structure, (2) presenting stimuli that help the students to retrieve information from the cognitive structure, (3) leading students to discover relationships by using cognitive-like inductive questions that help students to make conclusions, which is the new knowledge (see Table 12-1).

Table 12-1. Instructional Problem-solving Model

- Present problem based on real life. These must involve sensory modes.
- Analyze problem through discussion that helps students to elicit critical attributes of the concept; suggest a method for how this may be achieved. Suggest a hypothesis of these relationships.
- Introduce activities by manipulating materials and concrete objects relevant to the situation. Create a real-life model by retrieving relevant verbal information in the cognitive structure. Use inductive strategies such as questions. Student will make charts and models that help to report data.
- Deduce a generalization based on the relationships and summarize main points. Use questioning to help students activate their brains and create new knowledge. Refer to the original problem.
- Make a mathematical model or formula based on the generalization.
- Apply the new knowledge to other examples (reinforcement, transfer, and storage) in the cognitive structure so that materials will be stored in an organized way.

Guided practice will occur. Two sample problems follow.

Application of Instructional Problem-solving Model

The model for the two problems: (1) Area of Circle, (2) Linear Equation.

Problem Solving/Discovery

1. *Topic*: Area of Circle

Grade: 8
Time: 40 minutes

2. *Objectives*

a. Conceptual: To enable students to understand the that the area of any circle is $ЛR^2$.
b. Performance: Students will be able to calculate the area of a circle and to apply it correctly to calculating the area of the plywood in the given problem and other situations.

3. *Instructional Strategies*: Problem-solving/ discovery, cooperative groups and inductive questions.

4. *Materials and Equipment:* Plastic fraction squares and circles and paper circles.

5. *Procedures*

a. *Motivation*: Present the stimulus/problem:
An architect has been commissioned to build a church. The windows, which are round, have a diameter of fourteen feet. He wants a shade made of plywood to cover the window when it rains. The shade must fit exactly into the window space. There are a total of seventeen windows. How much plywood should he purchase for each window?

b. *Concept Development*

1. *Analysis of Problem*: Analyze the problem and note what needs to be done to find the area of a circular piece of wood. Retrieve the known facts in cognitive structure and write them on the chalkboard.
Circumference = D л or 2R л
Area of rectangle = L.W.
This will be done by inductive questioning.

2. *Suggested line of questioning*: What is the relationship between a circle and a rectangle? Use questions to deduce the

comparison of the area of a circle with that of a rectangle. How can this be found? Suggest a method. Rearrange the circle to make it a rectangle by cutting it into strips.

3. *Practical analysis of data*: Organize the class into cooperative groups. Each group will be given a task to cut the circle into strips and make it into a rectangle.

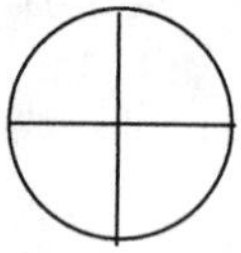

Students will manipulate the strips and fraction squares and note that the smaller the strips, the nearer the shape gets to a rectangle thus:

▼▼▼▼▼▼▼▼▼▼▼ Width or

Radius

▲▲▲▲▲▲▲▲▲▲▲

Length or ½ Circumference

4. *To create a real-life model*: The shape that has been created is a rectangle. Use questioning to deduce the following:

Circumference	= the 2 lengths of the rectangle
Length	= ½ circumference
Width	= the radius of the circle
Area	= ½ Circumference x Radius

5. *To make a generalization/mathematical model*: Combine the data in cognitive structure with the knowledge that has been developed.

Circumference $= 2R\pi$

Length $= \underline{½}$

$2R\pi$

$R\pi$

Width = Radius

Area of circle $= \pi R \times R$

Area of circle $= \pi R^2$

Note that the equation describes a real situation

6. *Application*: Apply new knowledge to the problem. To find the

amount of wood required for one window:
Diameter of window = 14 feet, R = 7 feet
Area of circle = ∏ 7 2

7. *Guided Practice*: There will be practice so that new knowledge is reinforced and organized into the cognitive structure. The class will discuss other situations to which the area of circle may apply. Students will verify the hypothesis by applying it to other situations. The class will calculate the area of a circle from an example given by the teacher.

8. *Independent Practice*: Groups of students will be given a project to complete at home. For example, you are making a number of circular clocks. Find how much glass you would need for a dozen clocks.

9. *Summary/Evaluation*: Students will create a grid for evaluating the project. The teacher will use performance evaluation. Projects will be displayed, and students will be required to explain their calculations and answers.

Real-life Problem-Situation

In the end, all mathematics must be applied to real-life problems or situations. The process begins by analyzing the situation, describing it in real-life terms, them translating the real-life equation into a mathematical equation. The process ends by applying the solution no the original problem. Table 12-2 presents the steps.

Table 12-2. Real-life Problem-Situation Model

- Analyze the situation.
- Create a real-life model by using letters that describe the situation. Example: total cost = TC.
- Translate the real-life model into mathematical model by using mathematical symbols. Example: X.
- Apply the mathematical solution to the original problem.

Application of Real-life Situation Instructional Model

1. *Topic*: Application of the slope and linear equation

2. *Objectives:*

   a. Conceptual: To enable students to understand that a linear equation involves a slope and an intercept.

   b. Performance: Students will be able to use the slope to write a linear equation and to compose a family telephone bill.

3. *Instructional Strategy*: Visual representations, questioning.

4. *Procedures*

   a. *Previous Knowledge*: To check the fact which student already know, the meaning of rate or slope, particularly positive, negative, and zero slopes. Prepare three charts to illustrate the weekly sale of videos for three artists.

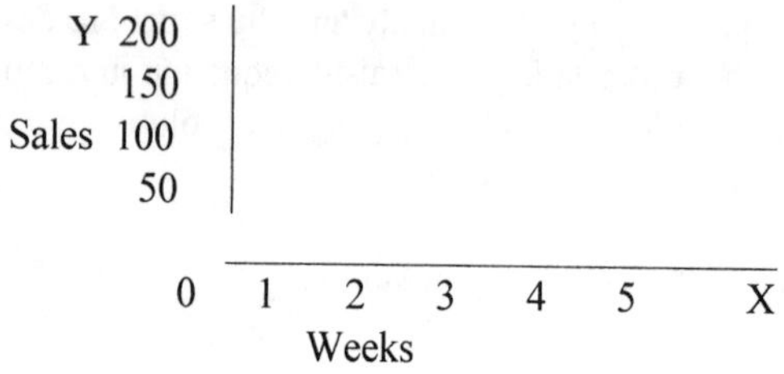

Review

(1) *How to write points on a graph*, e.g., (1, 50) (2, 100)
Students will notice that sales are described by a slope.
(2) *How to calculate the slope using a problem.*
Problem: John began to climb a hill. At 1:00 P.M. he was at 50 feet and at 5:00 P.M. he had climbed to 70 feet. Calculate the

$$\text{Rate of change or slope} = \frac{\text{Distance 2 - Distance 1}}{\text{Time 2 - Time 1}} = \frac{70 - 50}{5 - 1} = 5$$

Therefore slope is $\frac{\text{Change in Distance}}{\text{Change in Time}} = \frac{Y2 - Y1}{X2 - X1}$

Fact 1: To demonstrate that slope is on a straight line and the slope can be calculated from any two points.
Activity: Graph the line followed by John and calculate the slope.

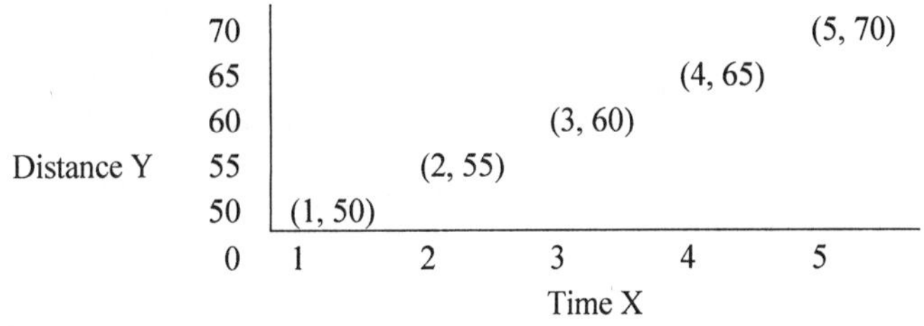

Rate or Slope $= \frac{70 - 50}{5 - 1} = 5$

b. *Motivation:* The Problem
John had travel. When he called his wife by telephone, the operator told him that the connection charge was ten cents and that the cost per call was five cents. Calculate the total cost of the telephone call.

c. *Developmental Activities*

Fact 1: Slope is used to calculate the cost of a single telephone call.
Activity: Student will suggest how to solve problem. Connection charge is 10 cents; rate is 5 cents. John speaks for 6 minutes. Students will prepare a chart describing the data, noting that at zero minutes the cost is 10 cents. Write on the chalkboard.

Cost in Cents	Time in minutes
40	6
35	5
30	4
25	3
20	2
15	1
10	0

The total cost will be 40 cents. Questioning will be the strategy.

Fact 2: An equation describes a real-life situation and may be written in real-life terms.
Activity: Teacher will use questioning to elicit an answer to how to write an equation. Students will write. Use questioning to apply order of operations.

Total Cost	=	Connection	+	(Rate per Minute	*	Additional Minutes)
40C	=	10C	+	(5C	*	6)
40	=	10	+	(5	*	6)
Y	=	A	+	B	*	X

The equation is C = 10 + 5X or C = 5X + 10 or Y = 5X + 10

Fact 3: A real-life equation may be written in mathematical terms.
Activity: Use questioning to deduce the formula Y = A + BX

Fact 4: The resulting equation is a straight line on a graph.
Activity: Students will use the data to write a graph. The cost will be represented by the Y axis with intervals of five to represent the rate of five cents or the slope. The time will be on the X axis with intervals of one minute. The slope represents the rate of change. Note that at zero minutes, the cost is ten cents. Cost per minute is five cents, then after one minute the cost is fifteen cents.

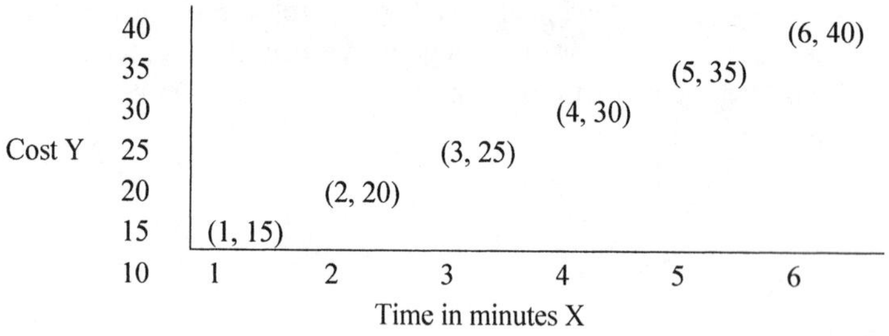

Fact 5: The graph and the equation must be interpreted.
Activity: Teacher will use questioning to get students to understand the graph.

Cost is represented on the		= Y Axis
Total Cost		= 40 Name
Minutes are represented on the		= X Axis
No of Additional Minutes		= 6 Name X
The graph intercepts the Y axis at		= 10 Name Y Intercept
Calculate the Slope	$\frac{40 - 10}{6 - 0}$	= 5 Name B
Real equation is	TC	= IC + (RxA)
	40	= 10 + (5x6)
Formula is	Y	= A + BX

Generalization: Equation to the straight line or *linear equation* is Y = A + BX.
A slope can be calculated from points on a graph.

A graph can be used to write an equation.
The equation can be used to describe and solve real-life problems.
The equation is the Slope-Intercept Form.
Generally, the formula is written Y = MX + B.
Activity: Use inductive questioning to translate real life equation in to formula and deduce the generalizations that will be written on CB.

Structured Practice: To ensure that the concept is organized in the cognitive structure, remembered, and prepared for transfer to other situations.

Activity: Organize students into groups or families. Each student will be a member who used the telephone. The group will compose the monthly telephone bill for the family. They will create a system of linear equations and graph them. Teacher will monitor progress. Finished products will be reported to the class.

Guided Practice: Teacher will ask students to use a situation to do a project.
(1) Given the population of five cities over a ten-year period, find the annual rate of change.
(2) The number of vacation tours taken by Americans increased by ten million per year between 1980 and 1990. In 1980 Americans took

300 vacation tours. Write an equation that describes the number of tours taken in 1989 and graph it. Hint: Make 1980 zero year.

5. *Summary/Evaluation:* Student will record the definition or formula of a linear equation and situations in which it may be used.

Concrete Representation

Empiricism is the foundation of mathematics; abstract concepts were derived from organizing concrete objects. This was so during the historical development of mathematics. Chief users were the Chinese who used turtles to express algebraic equations. For example, a trapezoid was represented by a "dustpan-shaped field," frustum by a "square field," and a tetrahedron by a turtle shoulder joint (Eves, 1969).

The Chinese still use concrete representation as a fundamental teaching strategy. James Stigler and Harold Stevenson (1991) conducted a comparative study of 120 classrooms in three countries: China, Japan, and the United States. The cities were Taipei (Taiwan), Sendai (Japan), and the Minneapolis metropolitan area in the United Sates. The authors found that each student in Sendai possessed a mathematics set of materials such as checkerboard, colored triangles, beads, and many other attractive objects. The authors described lessons in which the teachers used concrete objects to teach a concept. The teacher began with a problem, followed with analysis, and the conclusion was a definition or the formula. For example, the teacher wanted to teach fractions. In the classroom were jars and measuring cups. The teacher began with a problem, followed by analysis and a histogram. Students were only informed of the concept names at the end of the lesson after they understood the concepts.

Concrete representation is the tenth standard proposed by the National Council of Teacher of Mathematics (NCTM). It is a process standard or methodology standard. Concrete representation may be embodied in a conceptual development. In their nascent stages, concepts are derived from concrete realities; and from the concrete objects, abstract ideas are derived. In teaching a concept, the teacher needs to explore the criterial attributes of the concept. Concrete examples and nonexemplars are used to explicate the concept.

Gagne (1985) described the learning of concrete concepts as a human capacity that involved in putting things into a "class" and knowing the attributes of each member of class. Learning a new concept is incorporating new examples and attributes into the class that is in the memory or cognitive structure.

Conditions, such as capabilities in the learner, such as capabilities are a precondition for successfully learning of a concept. There must be a condition, such as verbal cues, within the situation. This includes presentation of stimuli such as instances of the concept. The learner learns to discriminate, analyze, and generalize. Table 12-3 provides an overview of the concrete representation instructional model.

Table 12-3. Concrete Representation Instructional Model

- Select some concrete item to represent the topic under review for example use squares and organize them represent factors in composite and prime numbers.
- Create a real-life representation of subject under review
- Use real-life terms to describe situation or problem example cardboard, paper or tangrams to represent squares, circles, or triangles.
- Manipulate the representations to create equations
- Transpose the real-life equations into mathematical equations.
- Close by referring to the original situation and to other problems.

Application of Concrete Representation Instructional Model

This section presents two plans (Quadratic Equation in Construction and Exponential Graph) that exemplify this method.

Plan 1

1. *Topic:* The Quadratic Equation in Construction

2. *Objectives:*
   a. *Conceptual*: To help students understand that the quadratic equation represents physical space, or area. The dimensions are the factors in the equation.

b. *Performance*: Students will be able to solve quadratic equations by using representation and by factoring.

3. *Instructional Strategy*: Constructing representations, inductive questioning

4. *Materials and Equipment*: Paper squares, tangrams, samples of telephone and electricity bills.

5. *Procedures*

a. *Motivation*: Present a situation in which a store wants to extend its show window to accommodate more display.

b. *Developmental Activities:*
Fact 1: An equation represents physical space
Activity: Present example of a store window
Length = X and Width = X
Students will cut a square piece of paper
Use questioning to emphasize that: Area = $X * X = X^2$

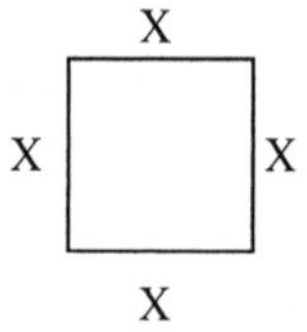

Fact 2: Additional space is also surface area. Equation may be described in physical terms.
Activity: Students will be required to cut seven strips of paper to represent the additional space to be added to the department store. Each is Length X and Width 1. They will arrange the strips thus:

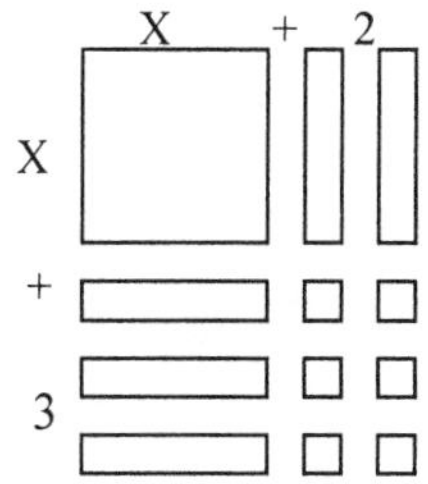

The new dimensions will be
Length = X + 5, Width = X + 3
Additional space will be filled in by 3 squares, each 1 x 1 on the length side and 2 squares of 1 x 1 on the width.
Area of additional space = 3 x 2 = 6
The new space is now a rectangle with area
A = the *square+5 units (each X * 1) + 6*
A = X^2 + 5X + 6

Students will verify the equation by counting the squares and emphasize that it represents physical space or area.

Generalization: The physical space is a square plus additions. It may be described as a mathematical equation called a quadratic equation.

Activity: Identify the dimensions and calculate area of the new rectangle:
L= X + 5, W = X + 3 are the factors of the equation, A= X^2 + 5X + 6 is a quadratic equation. Questioning will elicit the facts. Refer to several examples.

c. *Structured Practice*
1. Use physical method to find the area of space X^2+ 2X +3.
2. Factor the equation X^2 + 5X + 6

d. *Guided Practice*
Groups of students will be given projects (1) To find the area of given space and additions and write the quadratic equation and a corollary, and (2) To find the length and width by factoring the equation

e. *Independent Practice*
Student will be given assignments to factor quadratic equations.

6. *Summary/Evaluation*
The students will be given homework due at the next class meeting.

Plan 2

1. *Topic*: The Exponent
Grade: 8
Time 40 minutes

2. *Objectives:*
   a. Conceptual: Students will grasp the meaning of the concept of exponent and its effect on numbers.
   b. Performance: Students will be able to calculate exponents of numbers, graph exponent equations, and apply them.

3. *Procedures*

   a. *Motivation:* Students will be asked to examine bank statements based on simple and compound interests. They will see that compound interest is the more powerful instrument. This will create the reason why student need to understand how exponents function.

   b. *Developmental Activities*
   Fact 1: To establish the impact of exponents.
   Activity: Distribute sheets of letter-sized copy paper to each student. Direct them to perform the "folding paper experiment" and record the results thus:

Constant Fold	No. of Folds	Outcomes
2	1	2
2	2	4
2	3	8

2	4	16
2	5	32
2	6	64

Fact 2: To get students to observe that the outcomes represent patterns.

Activity: By questioning, lead students to write outcomes in another form:

2x2	=	4
2x2x2	=	8
2x2x2x2	=	16
2x2x2x2x2	=	32
2x2x2x2x2x2	=	64

Fact 3: The outcomes may be written in the form of equation.
Activity: Rewrite patterns thus:

4	=	2^2
8	=	2^3
16	=	2^4
32	=	2^5
64	=	2^6

Generalization: The outcomes may be written as $Y = B^x$.

Activity: Identify the parts of the equation and deduce their functions.
The outcome is (Y).
The constant 2 is called the Base (B).
The number of fold is the exponent (X).

Practice: Make a graph to test the power of the exponent.

X	B^x	Y
0	2^0	1
1	2^1	2
2	2^2	4
3	2^3	8
4	2^4	16

5	2^5	32
6	2^6	64
7	2^7	128
8	2^8	256
9	2^9	512
10	2^{10}	1024

Students will then draw the graph and notice the steepness of the slope. Students may make graph by using Excel.

Application: Consider the problem: In a certain specie of plant the number of cells is N and the number of gametes is determined by the equation $G = 2^{n.}$ Determine the number of gametes in a specie with ten cells.

4. *Summary/Evaluation*
Students will take a quiz on calculating exponents and will make use the Y function on the scientific calculator, then compare the exponential graph with a linear graph (see Figure 12-1).

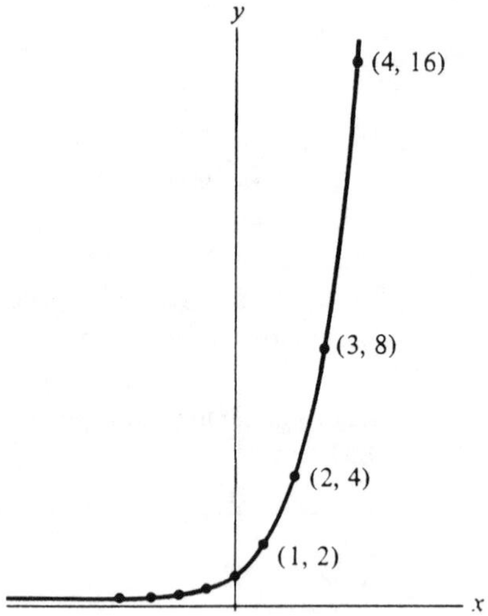

Figure 12-1. Computer Model of Problem Demonstrating the Power of Exponents

Concrete/Discovery

Concrete/discovery is based on the method of concept attainment. The model involves the shown in Table 12-4.

Table 12-4. Concrete /Discovery Model

- Presentation of examples: Teacher presents concrete examples of a concept and uses questioning to elicit attributes.
- Students are allowed to be actively involved in manipulating objects while teacher questions them to prompt examination of relationships and drawing of inferences.
- Students conduct experiments based on the concept. They gather data, record results, and make conclusions.
- The concept name is defined by making generalizations of observations. Students must be allowed to give information based on observations. Conclusions must be organized so that they may be organized in the memory and be ready for transfer to the next situation.
- Practice for reinforcement follows.

Application of Concrete/Discovery Model

1. *Topic*: Volume
   Grade: 8
   Time: 40 minutes

2. *Objectives*
   Conceptual: Students will learn the concept of volume is length, by width and height.
   Performance: At the end of the lesson, students will be able to calculate volume.

3. *Instructional Strategies:* Problem-solving, discovery, cooperative learning

4. *Materials/Equipment*: Plastic containers measuring 5 x 5 x 4 inches.

5. *Procedures*
   a. *Motivation:* Story of a manufacturer who wants to make glass jars to accommodate catfood pellets for cats. The dimensions of the jars must be 5 x 5 x 4 inches. He wants to make 100 jars. The manufacturer needs to know how many pellets will be contained in each jar and how much plastic to buy.

   b. *Concept Development:*
      1. *Analyze the problem*: Facts already known: concept of area, cube
      2. *Suggest practical solutions*: Experiment: Get a plastic container and measure the length, width and height. Get a number of cubes and confirm the dimensions 1 x 1 x 1 cubic inches. Fill the container with the cubes and count the total. Compare the dimensions of the jar and the number of pellets.
      3. *Arrange students in groups* and get them to repeat the experiment. Students will measure boxes, fill jars, and count cubes.

   c. *Generalization*: Use questioning to deduce mathematical; method of finding volume. Introduce concept name and formula Volume = L x W x H.

6. *Guided Practice:* Students will complete the problem above. A problem will be assigned to each group. Students will use Microsoft Excel to find volume.

Table 12-5 provides a comparison of methodologies as discussed above.

Table 12-5. Comparison of Subject-centered Traditional and Cognitive-Problem-solving-Neuroscience Methodologies

Traditional methodologies are based on authoritarian and nonauthoritarian methods derived from philosophy, whereas psychology provides methods that range from behaviorism to cognitive theories to neuroscience. Major principles derived from these theories may be summarized thus
Traditional and Behaviorist Methodologies
◦ Classroom is dominated by the teacher who makes the decisions. Students are given directions, work in silence, and get instant feedback. ◦ A statement of the objectives is written on the chalkboard. ◦ The teacher begins by giving the definition of the concept, followed by explanations and examples, and then by practice involving computation of problems and drill by students. The lesson proceeds in sequence of bits of information. ◦ The mind of the student is like a sponge, absorbing information. Students depend on the teacher for new knowledge.
Cognitive and Information Processing Methodologies
◦ This is a democratic classroom in which students have opportunities to participate and provide input into their learning. ◦ There is a problem-solving approach based on selection of realistic problems and students are given opportunities to discover new knowledge through meaningful activities and realistic resources. ◦ The lesson is suitably closed by students having input into the generalization of the concept, which is written on the chalkboard. Practice then follows; cooperative learning groups may be employed. ◦ Students are given opportunities to explore and the mind is actively engaged in creating new knowledge and in solving problems. ◦ The classroom is a busy and productive environment. Activities are used as the basis of gathering information and the emphasis is on understanding mathematics.

Chapter 13

The Learning Environment

Learning takes place in a social context. Within the content of the subject matter content, there are interactions among the teacher and students, students and students, and people and material resources. The dynamic mutual and reciprocal action and reaction of the individuals and groups creates a social situation that impacts on learning. Indeed the classroom is a stage, and all the individuals concerned are players, with roles and behaviors.

The classroom is a theater or arena in which the product is learning. The players or participants on this stage are the teacher, who is the leader, and the students, who are the players. The interaction between teacher and students, called *the teaching-learning process*, creates the social environment. The total action takes place within a physical environment, the classroom.

Leadership Style of the Teacher

The teacher is the legitimate leader in the classroom. This professional person was appointed as by a legal process and so has legitimate authority. This authority includes the power to organize the classroom, select textbooks, make arrangements on how learning takes place, and make all the decisions that make the classroom work. As the leader, it is the teacher who creates the vision for the future development of the classroom learning. The teacher is also the ultimate pedagogue in the classroom as has expert authority.

The leadership style of the teacher determines the classroom climate, including the decision-making process and the communication pattern. Leaders have different roles. The first is the vision for the class and decision-making regarding the direction in terms of the content to be taught, the pedagogical style, and how the resources should be applied. The teacher also decides on the rules of conduct for the classroom.

The Authoritarian Teacher

Leadership styles of the teacher have been described several ways. Kresh, Crutchflield, and Bellanchy (1962) described leadership on a continuum from authoritarian to democratic to laissez faire. Authoritarian leaders wield power. They make the decisions and are the ultimate judges and purveyors of justice. Communication is minimal and usually flows downward. As the group gets larger, the leader becomes more remote. Under the authoritarian leadership style, the teacher is in control of the class and demands cooperation, creates the competition, and makes all the decisions in the classroom including all the rules, what is learned, and when. Students in this classroom are generally well behaved and are generally silent; they listen passively, absorbing all the knowledge instead of using their minds to create new knowledge.

Many educators question the amount of learning that occurs in the authoritarian classroom. Some research has indicated that students remember more in the authoritarian setting; however, they are unable to transfer knowledge to other situations. Because transfer of knowledge is the essence of the learning process, authoritarian methodology has an inherent disadvantage.

The Democratic Teacher

A democratic leader exercises power by maximizing involvement and participation by students and spreads responsibility rather than allowing it to concentrate in the teacher's sole control. This spreads interpersonal relationships and seeks to reduce tension in furtherance of the goals of the group. By applying democratic principles, the teacher promotes cohesiveness and the achievement of the group. In a democratic classroom communication is open in all directions, that is, between teachers and students, and between student and students. This system encourages vertical and horizontal interaction Communication may be initiated by any member of the group, and content may be openly questioned. In this way learning is maximized.

In a democratic classroom, there is much participation. The teacher seeks cooperation, exacts responsibility, and gives encouragement. The teacher is open and friendly without being unprofessional. Students are able to work independently and creatively. There is less learning of facts than in the authoritarian classroom, but more learning of processes. The retention of facts may be less than in the authoritarian classroom, but there is more learning of processes so that students construct knowledge and meaning. Because students are able to discover knowledge for themselves, they are able to transfer this knowledge to new situations. This is the essence of learning.

Laissez faire means *anything goes*. In the laissez faire classroom, therefore, there is little discipline and less learning.

Leadership and the Motivation of the Teacher

McClelland (1988) described a model that proposed that the leadership style of teachers may be determined by their needs and motivations. The needs are (1) Need for Power (N-Power), where the teacher seeks power and therefore controls the classroom similar to the authoritarian leader; (2) Need for Achievement (N-Ach), where teachers are interested in their own achievement and that of the group. In this regard the teacher has vision and works toward clear goals. The social environment in this classroom is cordial and participatory, and learning is excellent. This is consistent with the democratic leadership; (3) Need for Affiliation (N-Aff), where the teacher is more concerned about a friendly social environment. This is similar to the laissez faire teacher. Nothing destroys the effectiveness of a teacher so quickly as allowing the relationship with students to become too friendly.

The Social Environment

The classroom is a miniature society. It includes diversity of human beings, and social relations are similar to those that occur in the larger society. The social environment of the classroom is as much a factor in learning as the teaching learning process. Researchers and professional educators agree on the power of the classroom climate to promote or inhibit learning. The social relationships in a classroom are called the *classroom climate.* A desirable classroom climate involves relationships of participation, cooperation, and understanding. In this social situation there are limits that everyone understands and respects including rules, purposes, and fairness. There are opportunities for the development of skills to provide competence and confidence. The purpose of learning is the cooperation and attainment of efficacious results. Maintaining classroom climate is one of the roles of the teacher.

Classroom Climate

Studies have been conducted on the influence of classroom climate on learning. Ned Flanders (1964) developed an instrument for measuring interaction in the classroom. Walberg and Anderson (1974) carried out research on the impact of learning environment on learning. They constructed an instrument called the *Learning Environment Inventory* (LEI), which had fifteen scales. Items included cohesiveness, democracy, difficulty, apathy, cliqueness, satisfaction, competition, goal direction, and favoritism. Responses were measured on a scale from strongly agree to strongly disagree. The questionnaire was administered to 1048 students in sixty-four classes in Montreal and the study found a positive relationship between classroom climate and learning.

Anderson (1972) did a study to ascertain whether learning environment could predict learning. He related the IQ to LEI on the cognitive and affective domains and behaviors measurements of student learning. LEI was responsible for thirty-six percent of learning outcomes, friction showed negative relationship, and IQ accounted for twenty-five percent. In conclusion, social environment affects learning; therefore, the teacher should maintain a classroom with positive climate.

Organizing the Classroom

The first principle of attaining a positive social climate in the classroom is the creation of an organized atmosphere. The physical environment must be prepared; the clerical work done; the textbooks, materials, and handouts available; and everything necessary must be prepared beforehand. Nothing causes disorder in the classroom more than the lack of preparation. If the teacher has to stop to check the mail or find a textbook, the flow of the material being presented is interrupted. Conversely, nothing fosters learning better than orderly procedures. The teacher needs to organize the furniture. Traditional methods are neat rows of desks. Alternative arrangements include creating clusters of desks for group work. The teacher should experiment with workable patterns.

In mathematics the teacher should seek to make the classroom into a laboratory. All equipment and resources should be laid out in geometrical shapes with rulers, manipulatives, overhead projectors, video tape recorders and so on in order. There should also be on display evidence of the subject, both ready-made articles such as the great mathematicians and models great inventions and group work.

The teacher must first know the name of every student. A plan of the seating arrangement of the class can be helpful in this. The seating plan is a list of names of all students prepared on an index card. It is placed on the edge of the desk for easy viewing. The teacher should use the seating plan to learn the names of students within two weeks of taking a new class by associating the names on the card with the faces of the students. The teacher should look at the card, select a name, match it with the student, and say the student's name. Student are often flattered that the teacher knows their names and will cooperate accordingly. Students know that they can evade and outwit a teacher's discipline if the teacher does not know their names. The teacher is well served to learn the name of every student.

The teacher should create for routines, such as how students enter and leave the room, and rules for answering questions. The teacher should insist that students do not answer in chorus, as this not only creates disorder but impedes learning. Many beginning teachers make the mistake of directing a question to the entire class without selecting a respondent. The teacher must insist that students raise their hands and must call on one student at a time.

Classroom Discipline

Discipline is the social and moral infrastructure of the school. (Brubacher, 1964). Learning cannot take place in chaos, it involves rules. Classroom discipline is the first rule for effective learning, or what lawyers call *conditions precedent.* It is the framework for instruction. Just as law and order are necessary for the proper functioning of the state, so discipline is important for the smooth operation of the classroom. Discipline is generally equated with order and is used to describe control and punishment in the classroom.

Discipline may be classified as extrinsic or intrinsic according to the methods used by the educator. Extrinsic includes the rules and the authority of the teacher. It is teacher-imposed discipline. Where teachers are in loco parentis, they are responsible for the conduct of the students. The teachers decide how they will wield authority in the classroom. With the authoritarian leadership style, the teacher is the legitimate authority in the classroom, and moral law requires that teachers make the rules and students simply obey them. The teacher is also responsible for punishment. The democratic teacher uses this authority more less overtly, as discipline will be coincident with and the outcome of interesting activities; the teacher's enthusiasm makes discipline intrinsic or self-discipline.

The next theory neither depends on the authority of the teacher nor the interests of the students. This relates to the child-centered classroom in which the children are freed from adult control and begin to take responsibility for their own actions. It involves the rule of the masses and based on the role of society in which social order is a requirement for the functioning of the school processes in order to achieve a purpose. Thus the children will discipline themselves, and each member of the group is responsible for his own conduct. Punishment is to protect the interest of the whole class, and so an offender becomes an example of the corrective measures.

The third theory is task-imposed discipline. When children are busy performing tasks that they enjoy, they discipline themselves.

Discipline as Value

Consider a person in possession of a golden ring. The ring has value because it is made of gold. The ring is valuable because the material from which it is made is socially valuable. The ring has extrinsic or instrumental value because of the material from which it is made. It can be sold for money. This value has a verb function.

The ring is also valuable to the possessor because it is a wedding ring that was given as a token of love and sentiment to the possessor. The ring has intrinsic value. The value has a noun function. Value or worthwhile behaviors may be either intrinsic or internal or extrinsic and external.

In the same way classroom discipline may be extrinsic, being imposed or conditioned by the teacher, or it may be intrinsic or internal, deriving its force from the internal motive power of the students themselves. Classroom discipline is essential because it determines the classroom climate and the quality of learning. *It is the moral infrastructure of the classroom.*

Models of Discipline

There are several models of classroom discipline based on the methods used by the teacher and the motivation of the students. The methods range from the authoritarian to the laissez faire based on the leadership style of the teacher, the classroom climate, and his philosophy of teaching.

There are several models of extrinsic discipline, one of which is behavior modification. Burrhus F. Skinner (1948) became something of a legend when he proposed the theory of operant conditioning that psychologists have applied to learning, motivation, and discipline. Skinner himself never proposed a model of classroom discipline. He was interested in learning. He conducted his experiments on hungry rats and from these experiments derived his laws rather than randomly. In operant conditioning the researcher controls the environment and decides what behaviors are considered acceptable. The researcher conditions the subjects accordingly. Like a law, the conditioned behavior may be generalized to other situations.

An operant is observable behavior, an act, or what Skinner called emitted behavior. It is voluntary behavior, that is, what a subject does, such as showing of hands. Skinner described operant conditioning as the probability of that response. Conditioning involves rewarding such behavior so that it will be more probable. By reinforcement, even complex behaviors can be shaped. Skinner himself conducted his experiment on learning. Other psychologists, such as Axerlrod (1977), and Ladoucer and Armstrong (1983), applied Skinner's methods to classroom discipline. Operant conditioning has been called Behavior Modification and is described thus:

1. The basic theory is that behavior is shaped by consequences; systematic use of rewards shapes behaviors.

2. Consequences determine whether or not the behavior will be repeated. Behaviors are strengthened if the consequences are positive and weakened if not.

3. Positive reinforcement increases the probability that the behavior will recur if repeated. If repetition follows the emission of a behavior, there is the probability that the act will recur.

4. In application behavior operates thus: The teacher sees a desired behavior and rewards it by giving praise. The student is likely to repeat it. The process of reward follows until the student is conditioned. Reinforcers are verbal and nonverbal. Verbal reinforcers include comments, grades. Nonverbal reinforcers are stars, facial expression, gestures, free time, prizes, and awards.

5. If the teacher observes an undesired act and ignores it, there is the probability that this undesired act will not recur.

Strategies of behavior modification include shaping, token economics, contracting and other methods described below.

Shaping occurs when the teacher decides on the target behavior and rewards each occurrence of it. The teacher checks for frequency and reinforces the behavior each time it occurs. For example, the student does not complete homework. The teacher decides on a number of problems that the student should do and a number of weeks during which the experiment should be conducted. When the student completes the worksheet the teacher rewards with praise, gives a grade, and exhibits the work the bulletin board. By the end of five weeks the students should have been conditioned.

Contingency Management, described by Charles and Barr (1989), includes other methods of behavior modification:

Token economics: The tokens are of no value, but the student may save them and exchange them for something of value after a period of time. Students who stays in their seats, raise their hands, and complete their assignments are rewarded with plastic chips that can be exchanged for toys, comic books, magazines, or other prizes at a later date.

Contracting: Contingent contracting is also a method of conditioning students. This is contract or agreement between the teacher and an individual student or a group of students to perform some act such as completing assignments. The behaviors are specified and so are the consequences. The contract may be verbal or written and for a specific period of time. A conditioned behavior will be generalized to other situations; for example, if students learn to show hands in mathematics class, they will also do so in English classes.

Rules-Ignore-Praise: In this approach the teacher formulates a set of rules to be followed by the class. These rules are published and understood and are then applied to the students through a system of conditioning. Thus positive response is praise and the negative response is ignore. This system gives credence to the adage that one "catches more flies with honey than with vinegar."

Many have criticized behavior modification as a classroom technique. Students soon discover that they are being bribed or that the token is not worthwhile. Some educators have raised questions of ethical behavior. Despite the questioning, behavior modification is an effective method of controlling students and effecting discipline in a classroom.

The Kounin Model

Kounin (1970) postulated that good discipline depends on good classroom management. He proposed that alertness, pacing, transitions, and individual accountancy are essential for smooth classroom functioning. The basic principles are as follows.

The Ripple Effect: The principle is that how the teacher corrects one student will have a ripple effect on the entire class. Kounin conducted experiments on students to demonstrate this principle.

"Withitness": Kounin proposed that the teacher should know what is happening in the class at all times. Even when the teacher's back is turned to the class as when writing on the chalkboard, the teacher "should have eyes at the back of his head." This is called withitness.

Transitions: Transitions between one activity and the next should be smooth, as this is critical to the effective classroom control. Any type of delay may create chaos in a few minutes.

Group Alertness: The teacher should hold each group accountable for the behavior of the entire group. Peer pressure is a certain way of exercising class control.

Assertive Discipline

Assertive Discipline is the theory by Lee Canter and his wife Marlene Canter (1987). Assertive discipline is based on communication style. Basically the three styles of communication are hostile, passive, and assertive.

An example would be a man goes into a bank and stands in a line. Another customer who has just entered the bank steps in front of him. He may respond in three ways: (1) hostile, such as being physically violent to the man; (2) passive, such as saying nothing until all the customers stand ahead of him and he is at the back of the line; or (3), assertively, such as saying, "Excuse me, this is not the end of the line."

Assertive discipline is very suitable for maintaining classroom discipline. It involves assertive speech. The steps are as follows.

Step 1. Recognizing and removing roadblocks to assertive discipline.

The expectations about students determine to a great extent how students behave and perform. If a teacher expects students to be bad, they will be bad. The teacher therefore needs to remove negative expectations that act as a roadblocks to good discipline. The teacher needs to:

Set limits, which are the rules or behaviors that are expected in the classroom. This should be done at the very beginning of the school year. The list of rules should be posted conspicuously. All students should be made aware of the rules, which should be explained to them. There should not be too many rules, and they should be discarded as they become obsolete.

Define expected behaviors, such as those expected from students, rewards for compliance, and punishment for noncompliance. These should be plainly stated, for example, "first infraction: loss of privilege; second infraction: conference with the principal; and third infraction: conference with parents." The new teacher should seek the cooperation of the principal and parents. Parent communication plays an important part in this process.

Step 2. Practice assertive responses.

The teacher should follow through with assertive communication. This is a kind of conditioning. The teacher should follow a rhythm of communication to ensure that the rules are observed.

Step 3. Be vigilant and consistent.

Canter (1987) stated that "no matter what the activity, in order to be assertive, one needs to be aware of what behavior you want and need from the students such as taking turns, not shouting out, starting to work on time, and listening when another student is speaking" (p. 62). The teacher should clearly instruct the students about these expected behaviors. The next step is to follow through on these expectations. Canter suggested the following verbal techniques:

Use hints, messages, questions, and demands to request appropriate behavior.

Deliver verbal limits such as tone of voice, eye contact, gestures, and use of students' names.

Use repetition to insist on certain behaviors

Step 4. Follow through on limits

Whenever there is a violation of a rule, the teacher should take action accordingly. Students will always seek to challenge the new teacher's authority and knowledge. At the first infraction of a problem, the teacher should there take corrective action. The teacher should

- Make promises, not threats.
- Decide beforehand on appropriate responses.
- Establish a sequence of enforceable behaviors and be firm and decisive.

Step 5. Implement a System of Positive Consequences

As in operant conditioning, behavior is determined by consequences. The teacher should initiate a system of positive consequences such as personal attention from the teacher (e.g., asking about the health of the student's mother, sending positive notes to parents or comments on a well done assignment, giving special rewards and privileges, and giving material rewards

The practice of assertive discipline should begin with the school year. The rules should be established to parents and the principal as well as the students. Rules should be enforced with firmness and decisiveness throughout the year. Assertive discipline is the most effective form of discipline.

Assertive discipline works every time. Students like order. They know that order is a precondition for learning. They will cooperate with the teacher in containing deviants. A teacher who is able to establish discipline at the beginning of the year will be able to maintain control of the class for the rest of the year. If a teacher fails to establish discipline at the very beginning, the class will exist in a state of uncertainty. Further if the teacher who has practiced laissez faire at the beginning of the year starts to assert authority midstream, students will regard him as unfair and learning in the classroom will be affected.

Intrinsic Discipline

In the end, intrinsic discipline means self-discipline. It operates by applying a series of constraints much as a dentist places braces on outgrowing teeth. With the establishment of limits and application of constraints, students become self-motivated and will discipline themselves. This should be the goal of all disciplinary exercise in a classroom. Models of intrinsic discipline have been proposed by Glaser (1986) and Driekurs and Cassel (1972).

Reality Therapy

This model by Glaser (1986) conceives of reality as a goal toward which students should be guided rather than focusing on matters from the past. The principles are that students can control their behavior, because they act rationally. The teacher's role is to help them to make good choices. The processes are

1. *Stress students' responsibility*: Students should be responsible for their own behavior. No excuses should be accepted; citing poor backgrounds is not good enough.

2. *Apply and enforce rules:* The teacher and students should decide on the goals of the class, set the rules and the consequences, and follow through to ensure that students make good choices to support the goal of becoming responsible citizens. Rules are the guide to success, not oppression.

3. *Provide suitable alternatives*: Because students are to choose their behaviors, they might select an inappropriate response. The teacher's role is to help them select alternative ways of behaving.

4. *Consequences*: Students should understand that every behavior carries consequences. Thus the teacher should follow through on consequences.

5. *Review*: Rules and behaviors should be reviewed constantly in class meetings and problem-solving sessions.

In effect this is a democratic method of discipline suitable for high school students. Real-life examples of what happens when bad choices are made will help students see the error of their ways.

Logical Consequences and Mistaken Goals

Discipline Without Tears by Rudolf Dreikurs (1972) is another method that advocates self-discipline. In operation, students must see that self-discipline is the goal and must be responsible for their own actions in the attainment of their goals. The aim is to help them see the consequences of misconduct in the real world, a process that should enable them to make better choices. Dreikurs does not equate discipline with punishment, which is he calls humiliation. Discipline is the freedom to make choices and understand the consequences. It requires that the teacher set limits in preparation for students' setting these limits for themselves. Dreikurs proposed the following steps in discipline.

1. Students are responsible for their actions. They must respect themselves and others and take responsibility for informing themselves of the rules and for knowing the consequences of misbehavior.

2. The important concept is that there are some mistaken goals that students must recognize, such as attention getting, power seeking, revenge seeking, and displaying inadequacy. To correct mistaken goals, the teacher should identify the mistaken goals, make students understand that it is undesirable behavior, inform them of logical consequences if such a goal is sought, and encourage students to seek good goals and to pursue them ardently.

3. Dreikurs suggests that the teacher is responsible for ensuring democratic discipline in the classroom by:

   - Providing leadership;
   - Setting limits;
   - Involving student in creating the rules;
   - Using firmness and kindness in maintaining the rules;
   - Inviting cooperation;
   - Allowing students freedom to explore, discover, and choose behavior; and
   - Promoting a sense of belonging in the group.

In effect, Dreikurs's model seems to suggest that misbehavior has unpleasant consequences and should be avoided at all costs. Dreikurs emphasized that misbehavior is associated with mistaken goals. Thus teachers should seek to identify these mistaken goals and encourage students to abandon these goals and seek good goals.

Causes of Disciplinary Problems

Disciplinary problems have been blamed on a multiplicity of factors from society to the home, the school, the students themselves, and especially to the classroom environment.

Society: In the past parents' support of the teacher was unconditional. A parent would say, "Teacher, I am leaving this student with you in your care." This is no longer the case. Parents still expect teacher to discipline their children but the social norms are different. There is an adversarial relationship between parents and schools in the matter of discipline, and parents are always willing to pose a challenge to the school. Since the advent of the students' rights revolution in the 1960s, the authority of the teacher has been undermined. Public opinion, parents, and the media, have all challenged the authority of the teacher. Public opinion needs to redefine the relationship between the home and the school and give more support to teachers. Teachers need to establish standards of conduct and ensure that they are above reproach.

The Classroom: Just as the social environment affects learning, so does the physical environment. The classroom furniture has to be suitably arranged, the lighting, temperature, and general cleanliness have to be conducive to learning. A disorderly classroom induces chaos.

The Teacher: The most important attribute of the teacher is preparedness. The teacher must be knowledgeable in the subject matter content and be skilled in pedagogy. This enables teachers to exercise the competence and confidence without which the lesson cannot be effective.

The Lesson Itself: The lesson has to be relevant to the needs of the students and society. It must be suitable to the academic level of the students, and have some purpose to their lives. A disorganized lesson will create indiscipline, therefore the teacher has to be well prepared and deliver the lesson with elegance.

Preventing Disciplinary Problems

Because discipline and order are the infrastructure on which rests the successful functioning of the classroom, prevention of problems should be the first priority. Some methods for preventing disciplinary problems are as follows.

Promote Good Student-teacher Relationships. Social relations are important in reducing conflict and promoting a positive environment. The teacher should recognize the achievements of the students, value good work, and reinforce good practices by verbal and nonverbal communication. The teacher should also seek to promote good relations by asking about their parents and their level of preparation for school.

Promote Self-discipline Among Students: Begin the year with the rules, expectations, and practices that should become the goad of self-discipline. The teacher should use strategies to reinforce good conduct until the standards are internalized into self discipline. To promote responsibility students should be taught to make choices and take responsibility for their actions. Also they should be given duties to perform. Set expectations for performance and evaluate their performance of the responsibilities and duties.

Promote Positive Social Relations: Teachers should know the names of their students. Questions should be addressed to individuals by name. Teachers should speak respectfully to the students and expect respect from them.

Promote Responsibility: Students should be given specific responsibilities for operating in the classroom. Duties such as organizing projects and conducting research may be assigned to students. They may also be given roles such as class leader.

Promote Fairness: Students like fairness. All students should be treated with fairness, without discrimination or favoritism.

Presentation of the Lesson: "Discipline is the outgrowth of productive activity in the classroom." If interesting work is taking place in the classroom discipline will naturally occur, and the teacher's role will be that of facilitator. The presentation of the lesson is critical in maintaining order in the classroom. Some strategies are

- Begin the lesson punctiliously,
- Give clear and unambiguous directions,
- Allow no slack time,

- Pace instruction to match the allocated time,
- Establish eye contact with students,
- Ensure that there are enough materials for every student,
- Address students by their names,
- Allow participation by all students, and
- Close every lesson.

Classroom Instructional Arrangements

The traditional classroom has a highly structured organized in with desks in neat rows and the teacher on a dais. This method is consistent with the authoritarian method. Modern classrooms have more flexible organizational styles. The teacher needs to have flexible classroom arrangements.

The Hidden Curriculum

The hidden curriculum consists of the practices in schools that influence the behaviors of students. Teachers help influence the hidden curriculum by implying what may be important in schools, such as behaviors, participation in activities, and interpersonal relationships. Some examples are the rules about contests and election or the informal agenda. It also includes norms and values. These are found in textbooks, for example, in the form of problems that are in the mathematics textbooks. The organizational structure of the individual classroom also sends messages to students, especially the nature of the interaction between teachers and students.

Modern methods for organizing a classroom include individualized instruction, independent learning, group instruction, and cooperative learning.

Individualized instruction is a laudable but difficult method of instruction that is best used with disabled or gifted students. In this case the teacher acts as a resource for coping with individual differences in a classroom and as a guide and monitor of students' progress rather than providing formal instruction as in a traditional classroom. Computer resources are very helpful in for individualized instruction.

Independent learning is also individualized. Topics are negotiated between teacher and students. Students assume personal responsibility for the assignment, and the role of the teacher is that of facilitator, consultant, adviser, and evaluator of the students' work. Portfolios are suitable for this type of instruction

In group instruction the entire class may be a group, or the teacher may organize the class into small groups. The basic reason for arranging students in this fashion is to accommodate their interests and needs. The teacher's role is to help students to plan, coordinate and monitor the groups. The group teaches itself.

Cooperative Learning Groups

Today's corporate world relies on group activity to attain objectives. Cooperative learning groups prepare students for the corporate world. The groups foster cooperation, a necessary social skill, rather than competition. Cooperative groups are usually heterogeneous, and students have the opportunity to learn from their peers. Four models of cooperative groups have been proposed:

1. The structured approach (Kagan, 1990) involves creating pairs of students to list vocabulary words to the level of mastery. Many students relegate this to the level of fun.
2. The group investigation approach (Sharon and Sharon, 1994) allows students to work together to find answers to topics in which they are interested. Each student has a task to find specific piece of information, which is then brought to the group for synthesis. This is a problem-solving-inquiry approach. The question is posed, and the students brainstorm for a hypothesis. They then divide the group into groups to find aspects of the answer and the group then evaluates the work. Cohen (1999) suggested that each individual has different abilities. The investigation group helps individual students to use their abilities. With the cooperative group each student is given an assignment geared to that student's particular level of ability, which promotes effective performance at that level.
3. The curriculum package group by (Slavin, 1995) is based on curriculum packages that have cooperative learning structured into the material. Slavin (1989) proposed a strategy in which the group has a goal but with individual accountability. Orlick, Harder, Callahan and Gibson (1988) suggested that the teacher still has the responsibility for setting the stage and working with students while students work among themselves.

4. The jigsaw strategy (Johnson, Johnson & Holubec, 1994) depends on experts. The subject is divided into segments. Assignments are given to individuals in their home groups. These experts then migrate from their groups and aggregate in an expert group. The students become experts in one area of the subject, after which the experts return to their home group, where they teach the other members of the group the topic in which they have become experts. Each member is dependent on the other for learning. If the expert fails, then the whole group of students fails. The expert has to be responsible. Individual students are responsible for their own performance such as on a test. They are also responsible for the performance of the group as a group. In this case standards for performance have to be established beforehand.

Conclusion

Like adults, students have needs that can be satisfied by working in groups. The need for power, attention, and affiliation are best satisfied through an interactive classroom setting and by working in teams. The process of getting satisfaction promotes intrinsic motivation and inspires students to learn more. As learning becomes a motive force, it directs behavior and interest in learning and improves academic performance (Glaser, 1986).

ASSIGNMENT

1. Evaluate two models of discipline and say how you would implement them in the classroom.

2. Describe the leadership styles of teachers.

3. Evaluate the importance of the social environment in teaching and learning.

4. Describe ways in which you may create a productive mathematics environment.

5. Imagine that you are a beginning teacher. Describe how you would establish discipline in your classroom.

6. What is the hidden curriculum? Describe the impact of the hidden curriculum in the classroom environment.

7. Describe some methods by which a teacher may organize a classroom and evaluate the relative impact of each method.

Chapter 14

The Mathematics Curriculum

A curriculum describes the ground that students and teachers cover to reach the goal of education. The Latin origin describes it as "runway" or the course that one must traverse, as in a race (Brubacher, 1964). Curriculum may be described in two major ways: (1) as a program of studies or a collection of courses at a specific educational level, or (2) in a broader view, as all of the experiences of a child or youth in educational institutions (Travers & Rebore, 1995). With regard to the latter point, the curriculum includes the extra curricular activities, such a mathematics club, mathematics fair, and the like.

Curriculum is based on the educational aims with three functions. First they give direction to the process of education. It is generally acknowledged that aims Goals enable planning; without them, any program is like a rudderless ship. Second, goals are motivating. They represent the values that will make educators want to attain objectives. Third, the aims represent the criteria by which programs may be evaluated. The school knows how far it has progressed by comparing the goals of the curriculum against the performance.

The goals of a program are based on philosophy, culture, and the needs of society and the learners themselves. Philosophy gives direction to the goals of education and the pedagogy or the methods by which the educators attain those goals. It also gives direction to classroom discipline. Moral education is value-laden.

Philosophy analyzed the place of morality in discipline, but in the end it is the law that determines how discipline is administered.

Curriculum content is derived from many sources, such as culture. Taba (1992) suggested that these sources are culture and society. The chief contributor is culture, which determines the content and is the accumulation of experiences of the race called the cultural heritage. It also includes present situations. Some researchers have claimed that the only purpose of the cultural heritage is to prepare the young to live in the present time.

The main purpose of education is to equip the young to live in the present world by providing them with the skills necessary for survival. Education prepares students with technological skills for the society. Hence the workplace determines and the level technological development dictates what should be taught in schools. Recent ideologies have suggested that pedagogy should be based on the learners' reality, that is, learning activities should take their cue from activities in the real world so that students may link education with reality.

Psychology provides the learning theories that guide the teachers' planning of pedagogical activities. The goal of education is learning and teaching is most efficient when it conforms to the natural tendencies of the learners. Theories of learning and theories of motivation guide the teacher into practicing good pedagogy and facilitate student learning.

Tyler (1993) suggested that curriculum objectives should be derived from the learners themselves, contemporary life outside the school, the subject matter itself, philosophy, and psychological theories of learning. The curriculum is derived from several sources that are all interconnected. The end product is devolved into the teaching-learning process in the classroom. The state Department of Education takes all of these curriculum purposes, philosophies, and cultural and subject matter content into consideration and organizes them into a usable curriculum.

The Curriculum and Society

Dewey (1938) suggested that the curriculum reflects society, which may be either authoritarian or democratic. The curriculum that instills in students the social heritage, including knowledge, attitudes, and values.

Dewey suggested that the primary objective is to prepare students to face responsibilities and manage success later in life by helping them to absorb organized bodies of information and acquire skills. Subject matter and standards of conduct are legacies from past generations; to absorb these, students must be docile, receptive, and obedient. Textbooks are the repositories of wisdom and teachers are the agents that provide the connection. Dewey was describing a curriculum in which children learn the attitudes of the past, which are irrelevant to the present; they become dependent on teachers and textbooks. Such a curriculum is geared toward maintaining an elitist education system and discriminating against the masses. It is consistent with subject-centered curriculum and based on authoritarian pedagogy.

The other curriculum model also described by Dewey is child-centered, based on social needs, and the experiences of children. It promotes democratic values. It encourages students to be independent, cooperative, and creative because its methods and objectives are of the present time. Dewey believed that the authoritarian style hinders the development of individuality, external discipline, and learning both from books and teachers or by experience. Students are more likely to acquire specific skills when the study involves attaining skills that have some direct relationship to real life in a changing world.

Indeed this is the very essence of a curriculum that trains individual students for a changing and dynamic society. The following historical account of curriculum in the United States demonstrates that the curriculum was created and adapted for a changing society. Evolving from the academic, subject-centered curriculum with mental discipline as the aim, the curriculum has become more utilitarian and standards-based.

The History of Goals of Mathematics and Society

As society changed and become more complex and sophisticated over the centuries, so too did the goals of mathematics. Mathematics was invented to serve social and economic goals, hence there was always need to change the goals. The goals of mathematics, however, have not always been consistent with social goals, nor have they kept pace accordingly.

Throughout the history of the United States, the needs of the society have always shaped the goals of education. During the colonial era, the primary purpose of education was religious and so was the curriculum. Education was for the elite; hence the curriculum was based on classical curriculum inherited from Europe. High school education was given in the Latin grammar schools and paralleled the education of those who were sent to England for schooling. Industrialization of the nineteenth century prompted new demand for a utilitarian education to fuel the economy. There was greater demand for a curriculum that included mathematics. In the nineteenth century education became more widespread with the founding of the first public high school in 1821, and the Kalamazoo Decision of 1874, which authorized the use of public funds for schooling at the secondary level. Latin and Greek lost their stranglehold as a medium of instruction, and new subjects entered the curriculum. Mathematics became important to fuel the growing industrial economy. Great European philosophers such as Friedreich Froebel, Johann Pestalozzi, and Friedrich Herbart influenced pedagogy in this period.

Significant curriculum reforms began in the late nineteenth century with the formation of the National Education Association and the first *Committee of Fifteen* in 1890 and then the *Committee of Ten* in 1893. These committees were tasked with recommending curricular reform and produced the *Carnegie Report* of 1910. The Committee of Ten Report (1893) was written under the chairmanship of Charles Elliot, then President of Harvard University (Hebowitsh & Tellez, 1997). The report outlined clearly what should be taught in high schools and set a standard for others to follow. It supported the doctrine of mental discipline to be taught uniformly through nine subjects as a means of training the mind (see Tables 14-1 and 14-2 for curriculum recommendations).

The NEA also sponsored *A Committee of Fifteen* in 1895, under chairman William Torey, Commissioner of Education. This committee stressed the liberal arts curriculum. In 1913, however, with the publication of the *Cardinal Principles and Reorganization of Secondary Schools*, a new era began. At that time World War I brought the U.S. into relationships with other nations. This era also saw mass immigration, especially from Eastern and Southern Europe. Further, technology entered the scene in this era. The first computer was invented in 1945 and gave rise to the establishment of the National Science Foundation. Burgeoning ethnic diversity gave schools a new mission, and the advent of technology created new knowledge.

Table 14-1. Foundations of Curriculum and Lesson Planning Model

PHILOSOPHY Aims of Education Epistemology Axiology Technological Change	⟷	SOCIETY Reproduction of The Status Quo Social Change Globalization
^		^
CULTURE Ideology Cultural Heritage	⟷	PSYCHOLOGY Learning Theories Motivation
↑		↑
	THE STATE Learning Standards	
	↑	
	THE SCHOOL	
	↑	↑
SCHOOL SUPPORT Community Equipment, Materials Finance Schedule	CURRICULUM Knowledge of Subject Principles of Instruction Learning Sequences	
	↑	
	THE INSTRUCTIONAL PROCESS Objectives Content Activities and Motivation Evaluation	
	↑	
	LEARNING OUTCOMES Cognitive Affective	

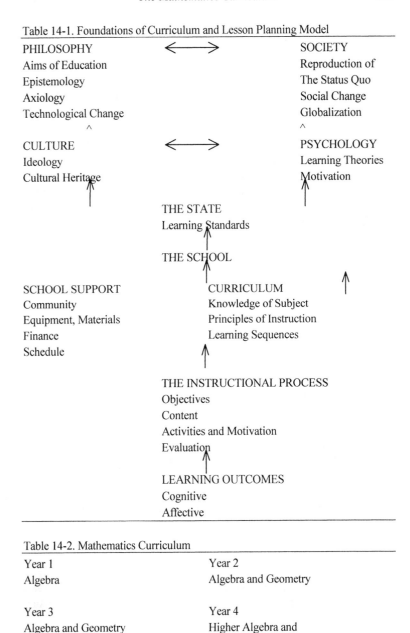

Table 14-2. Mathematics Curriculum

Year 1 Algebra	Year 2 Algebra and Geometry
Year 3 Algebra and Geometry	Year 4 Higher Algebra and Trigonometry

All these influences demand that the public school provide a unifying principle and a new purpose. Harris (1983) stated that the new directions were reflected in the Seven Cardinal Principles, which shifted the direction of the high school from college preparatory to comprehensive. The report stated that education should be reflect the needs of society.

This report was followed by an *Eight-Year Study* for the period 1933 to 1940. This study was the work of the committees that had shaped the structure of the curriculum during that period. The Carnegie Committee, for example, gave to the curriculum the Carnegie unit, which awarded "a class meeting of five periods weekly for a total of 120 clock hours per school year" (p. 118). The emphasis was on language and mathematics. Foreign language was given two units and mathematics two units more, whereas other subjects were awarded one or one-half unit (Boyer, 1983).

One aspect of this influence was the Life Adjustment ideology, which was implemented into the curriculum of schools. Life Adjustment was first conceived in 1945 when the U.S. Office of Education sponsored a conference on vocational education at which the term *life adjustment* was first coined. Life adjustment has been not successful in defining education. In fact, in 1957 when Sputnik dramatically put the United States behind Russia in space program, life adjustment education was regarded by some as the culprit (Travers & Rebore, 1995). To that time the structure of mathematics curriculum had been traditional and had included algebra, geometry, and trigonometry.

The implementation of the policy occasioned by the National Defense Act of 1958 was dramatically altered the curriculum. The emphasis on algebra, geometry, and trigonometry as discrete units was to be changed.

National Influences on the Mathematics Curriculum

The school curriculum in the United States has always been implemented in the context of social, political, economic, and national concerns. Chief among these curricular areas has been mathematics, which has received particular emphasis during then twentieth century.

The launch of Sputnik in 1957 stimulated the development of intensive study in mathematics, science, and languages in schools all over America. Some regarded the United States as inferior to Russia in the area of technology. The remedy was to reform the mathematics curriculum, which prompted the rewriting of the curriculum during the 1960s. Another group that advocated reform during the Sputnik era was the Association for Supervision and Curriculum Development, which suggested innovative projects in curriculum. The Council for Basic Education proposed that standards be used for measuring curriculum outcomes, and the NEA advocated curriculum reform up to grade twelve. In all, these reforms affected the curriculum and content began to change. For example, the Sputnik age ushered in the "new mathematics," including non-Euclidian mathematics, which had previously been reserved for the college curriculum was then introduced into the high school curriculum.

In 1960 the National Science Foundation was instrumental in initiating reform in mathematics. Content was viewed in terms of its structure and was pushed downward into the lower grades. Consequently, some students may now be exposed earlier to mathematical areas including algebra, calçulus, and statistics plus commonplace themes such as discovery and reasoning.

National Reports

National Reports have had dramatic influences on curriculum reform. Chief among these was the *Nation at Risk,* published in 1983 promulgated by the National Commission on Excellence in Education. This report triggered genuine reform at both the national and state levels. Among these reforms were new methodologies, which specified objectives for learning, mastery in learning and teaching, and standards for effective testing.

The Report on Excellence proposed a core curriculum with eleven subjects including mathematics. The report suggested that "in high schools, all students should expand their capacities to think quantitatively and to make intelligent decisions regarding situations involving measurable qualities. Specifically, we believe that all high schools should require a two year mathematics sequence for graduation and that additional courses are provided for students who qualify to take them" (Boyer, 1983, p. 71).

In implementation, the reforms were very effective. The *Condition of Education*, published by the U.S. Department of Education, gave data to show that there were higher standards for graduation, better academic performance, and graduation rates. Indeed the status of high schools was elevated, and they received better public recognition.

In 1994 Congress passed Goals 2000, which encouraged states to work together to develop curricula for students. Goals 2000 is viewed by many as the first step in developing national standards, a position that has traditionally been vehemently rejected by policymakers since the establishment of the Union.

Professional Organizations

In the 1980 another national body greatly influenced the mathematics curriculum. This was the National Council of Teachers of Mathematics (NCTM). During the 1970s, performance of students was perceived to decline. In 1989, NCTM introduced the national standards for mathematics. The were revised in 1995 and again in 2000. These standards govern three levels (grades K–4, 5–8 and 7–12) in both mathematics content and in process (see Tables 14-3, 14-4, and 14-5.

State Influences on the Curriculum

In the United States education is generally the responsibility of the individual states. The state constitution determines the goals and direction of education in each state. Although the federal government has become involved in education whenever national security and policy issues arise, it is the states that bear the burden of education direction? Generally there is coordination between the federal government and the states. For example, when the national standards in mathematics were accepted by the federal government, the states used it as a pattern on which to build their individual state standards. Power for educational direction and control was given through commission to the states by the Tenth Amendment which states, "The power not delegated to the United States by the Constitution, not prohibited by it to the States are reserved to the States respectively and to the people."

The state legislature and the state Department of Education develop curriculum goals, content, and evaluation standards. The curricula are implemented by the local school districts, which have power given to them by the state. Historically, the State Department published curriculum guidelines to assist teachers in implementing the curriculum.

Table 14-3. The 1989 NCTM Curriculum Standards for Grades 5–8

1. Mathematics as Problem Solving
2. Mathematics as Communication
3. Mathematics as Reasoning
4. Mathematics as Connection
5. Number and Number Relationships
6. Number Systems and Number Theory
7. Computation and Estimation
8. Pattern and function
9. Algebra
10. Statistics
11. Probability
12. Geometry
13. Measurement

Table 14-4. The 1989 NCTM Curriculum Standards for Grades 9–12

1. Mathematics as Problem–solving
2. Mathematics as Communication
3. Mathematics as Reasoning
4. Mathematics as Connection
5. Algebra
6. Functions
7. Geometry
8. Geometry as Synthetic Perspective
9. Geometry as Algebraic Perspective
10. Trigonometry
11. Statistics
12. Probability
13. Discrete Mathematics
14. Conceptual Underpinnings of Calculus
15. Mathematical Structure

Table 14-5. 2000 NCTM Curriculum Standards: PreK–12

A. *Mathematics Standards*
1. Number and Operation
2. Patterns, Function, and Algebra
3. Geometry and Spatial Sense
4. Measurement
5. Data Analysis, Statistics, and Probability

B. *Process Standards*
6. Problem-solving
7. Reasoning and Proof
8. Communication
9. Connection
10. Representation

The guides specify the goals, objectives, competencies, and instructional activities for each grade level. Of course teachers supplement these guidelines with textbooks and materials of their own.

During the 1990s, the states followed the lead of the federal government in setting standards for each subject including mathematics. In New York, mathematics is integrated with science and technology and is called Mathematics, Science, and Technology. New York State standards are shown in Table 14-6 as abbreviated standards on which schools are required to base their curricula.

Table 14-6. New York State Standards for Mathematics

Standard 1
Students will use mathematical analysis, scientific inquiry, and engineering design, as appropriate, to pose questions, seeks answers, and develop solutions.
Standard 2
Students will access, generate, process, and transfer information using appropriate technologies.
Standard 7
Students will apply the knowledge and thinking skills of mathematics, science, and technology to address real life problems and make informed decisions.

Curriculum Models

Historically students attending high schools in New York obtained one of two different levels of certification based on the curriculum pursued. First there was the Regents-endorsed certificate for those students who were college-bound. This curriculum was academic and geared for students who planned to attend college. To obtain a Regents-endorsed diploma, students had to pursue a mathematics curriculum called Sequential I, II, and III. Second, the majority of students who were not college-bound were differentiated into vocational and business tracks. The majority of students took the Regents Competency Examination (RCT), which was a less rigorous curriculum not geared for college study. Thus the curriculum was double tracked for the elite and the others.

In 1998 the state decided that all students should take the Regents examination, including both college-bound and non-college-bound students. The curriculum therefore consists of a core curriculum divided into two syllabuses: A and B. In reality the content of Sequential I, II, and III has been reorganized into two tracks. More important, some topics, especially the more difficult ones have been removed from the content, so that each course is now less rigorous than in the past.

Curriculum Organization

There are many ways of designing a curriculum, but the subject-centered and child-centered structures are the best known. Curriculum may be described on a continuum from subject-centered to student-centered based on philosophical positions. Table 14-7 below describes the curricula.

Table 14-7. Curriculum Continuum from Subject- to Student-centered

Subject-Centered			Student-Centered
Separate Courses for Each Subject Discipline	Integrated or Fused	Core or Basic Programs	Activity

The Subject-centered Curriculum

The subject-centered curriculum is most widely known and practiced. It is based on Idealist and Realist philosophies and on Essentialist learning theory in that it proposes that there is a body of knowledge that is essential to the survival of the society, like the Great Books curriculum. Students study single subjects that are discrete and unrelated to any other subject in the curriculum. The subjects are the traditional Liberal Arts inherited from the Romans. Boethus first coined the terms *trivium* (grammar, rhetoric, and logic) and *quadrivium* (arithmetic, geometry, music, and astronomy) to encompass the Seven Liberal Arts. Although subjects have been added to the curriculum, the basic structure of the Seven Liberal Arts remains in place.

The Integrated Curriculum

The aim is to combine subjects so that they become meaningful to students. It is mostly used at the elementary level to reduce the number of subjects taken by students. Integrated content includes English/ Language Arts, which consists of reading, writing, spelling, grammar, speech, and literature.

The Core Curriculum

The fused or broad-fields curriculum, as it was originally named in the 1930s, was conceived as interdisciplinary (Webb & Jordan, 2000). In the 1980s, it has became regarded as the comprehensive body of knowledge that is universally required in a technological society. The aim is to promote problem-solving in a model that stresses facts, principles, socioeconomic conditions, and moral values. It is based on Perennialist philosophy and is near to the extreme of the form of student-centered curriculum. The Mathematics, Science, and Technology Learning Standards promulgated by New York State is an example of a core curriculum.

The Activity Curriculum

The Activity Curriculum is also called Constructivism. It purports that the child is the pure center of learning. Activities are based on the child's experiences and embrace all subject areas.

The School Program

Mathematics operates as a mixture of the subject-centered and core curriculum. Since the 1980s students in New York have obtained different levels of certification based on the curriculum they pursued. First there was the Regents-endorsed certification for college-bound students. Students pursued an academic curriculum and took an examination based on the standards set by the Board of Regents. The majority of students pursued a less rigorous course of study and took the Regents Competency Test (RCT), which was not geared for college, at different levels. Such students were obliged to take remedial courses at the eighth grade level if they were admitted to college.

In the 1990s, with the emphasis on quality education and standards, the Regents changed its policy. All students are now required to take the Regents Examinations as a condition for graduation from college. Accordingly the schools have reorganized the mathematics programs. Sequential I, II, and III were replaced by Mathematics A and B. Curriculum models give direction to methods of pedagogy ranging from authoritarian to nonauthoritarian.

Conclusion

The curriculum, based on the cultural heritage in contemporary life, has a life of its own. It is the medium by which numbers are made to solve everyday problems. The main goal of every curriculum is to ensure that students learn, and an appropriate curriculum best achieves the goal.

Philosophy, psychology, culture, and society are among the influences on the curriculum, but it is the state that has the legitimate responsibility for coordinating all these influences and creating a viable, dynamic, and operational curriculum. Current organizations such as the NCTM also exert some influence on the curriculum content. Teachers are the agents for implementing these curricula, and their knowledge and expertise govern the effectiveness of the curriculum. It is therefore important that teachers understand the relationship between society and mathematics and plan the appropriate content to attain its goals and for ensuring that the curriculum becomes the agent for transforming individual lives and the society which is the cradle of civilization.

ASSIGNMENT

1. Read the NCTM standards and compares it with the State standards in mathematics.

2. In New York, the mathematics curriculum has been revised. The structure Sequential II and III has been replaced by a new arrangement called Syllabus A and B. Examine and evaluate the content of these curricula.

3. Make a list of the main topics in syllabus A and B and select suitable textbooks that may be used to teach them.

4. Evaluate the influences on the curricula of the following: philosophy, culture, psychology, and society.

5. Describe the role of the state in curriculum development.

6. Describe the work of committees on the mathematics curriculum.

7. What are the differences between the integrated, activity, and subject-centered curricula?

8. Choose one teaching unit and write performance indicators for the unit.

Figure 14-1. The Tortuguero Canal, Costa Rica

Chapter 15

Testing and Measurement in Mathematics

Laymen and educators have always complained about the effectiveness of tests in determining the achievement of students. To date, however, no-one has proposed a better method of evaluating performance. Despite their shortcomings, in whatever form they are administered, tests exist for the purpose of determining the achievement of students in a given subject. They are also used to predict future performance of students in specified areas such as academic. Tests and the curriculum content are complementary in determining the outcomes of the educational process and the preparedness of students to fulfill their roles in society. Tests may be classified by method by which they are constructed, by purpose or intention of the tester, and by types such as achievement.

Classification by Method

Some tests are defined by their method of construction or the design by which they are made. These descriptions are reported in dichotomy such as group or individual, power and speed, objective or subjective, and performance and pen-and-pencil tests.

Group tests are given to a representative sample or group of students such as a class. They measure the performance of the individual, and the result ranks the student in relation to others of the same group, thereby comparing one student with another. Usually the teacher analyzes the results of the test statistically. For example, the teacher finds the mean of performance of the group and is able to compare the relative performance of each student above or below the mean. Group tests enable the testing of large number of children by a nonprofessional psychologist. The test is empirical, evaluating the extent to which student has mastered the specified facts.

Individual tests on the other hand are test the progress of individual students from one period to another, so as to facilitate evaluation. Individual tests are usually done for home-schooled students and for programmed instruction.

Power tests measure performance based on items that involve an arrangement of items in increasing levels of difficulty. These tests have no time limits and take into account differences in performance as well as the complexity and refinement of thought of which persons are capable rather than the pace at which they complete the test (Lewis, 1974). The items are arranged in ascending order of difficulty. Everyone can answer the easy questions that appear at the beginning of the test. The test selects the students, so that few can answer the last questions. It is a power test because students work up the difficulty gradient until they are screened out their individual limits.

Speed tests are based on speed. All questions on a speed test are at the same level of difficulty, and all items can be answered correctly by all students given sufficient time. In practice all tests are a mixture of power and speed, with a mixture of difficulty gradient.

A *standardized test* is one that has been standardized, meaning it was administered to a representative group of a designated population of such persons classified with respect to age, gender, socioeconomic class, or geographical location. Standardize tests are standardized by administering them to sample groups for which the test is intended. The scores are analyzed statistically for the mean or norm that may be generalized to the target population. This test norm represents the average performance of the group and may be used to evaluate the performance of others in the same population at some future date.

Teacher-made tests are created by the teacher and are criterion-referenced, that is, based on the objective and content of the particular course.

An *objective test* is one that can be made objectively, without the judgment of anyone. Different examiners who mark the test should get the same results. The correct answers and the marking scale are determined beforehand, so that the incidence of errors can be zero. A *subjective test* on the other hand is largely based on the judgment of the examiner. In reality, the examiner determines the answers beforehand, but there are some characteristics that will be judged according to the subjective impression of the examine, such as the organization of a presentation. Examples of subjective tests are essays, which are evaluated according to the judgment of professionals.

A performance test, as the name implies, is based on the performance of some skill such as a project or display. A paper-and-pencil test is a written test.

Classification by Purpose

An *achievement test* is designed to assess achievement in a particular field, such as a school subject. They are usually criterion-referenced and based on the content of the course. An example of an achievement test is the Regents examination. Mental Ability Tests evaluate the ability or intelligence of the individual at a particular age group. They are generally IQ tests.

An *aptitude test* measures the potential of an individual for achieving in a particular subject, such as mathematics. They are a combination of ability and attainment tests. They are based on knowledge of content and on verbal and mathematical skills, but they are not designed as a particular test of content. Aptitude tests aid in selecting the best students for higher education and are generally based on higher-order levels of thinking and knowledge. For example, the scholastic aptitude test in mathematics is more geared to knowledge gained during schooling at the upper secondary level. According to Lewis (1974), these tests seek to assess a student's ingenuity in dealing with novel situations as well as manipulative skills, and specialized areas of mathematical skills such as numerical and spatial reasoning. Other types of tests are *personality tests* and tests of *divergent thinking*.

Classification by Types

A *norm-referenced test* is designed to evaluate the average ability of a group of students. It measures the typical performance of individuals in a given category such as age, or grade level. Because typical performance is stated as the average of a group, norm referenced tests are standardized.

Criterion-referenced tests measure the performance of individuals compared to others in the same group based on instructional objectives of the course. Items are selected from the course content in accordance with specifications rostered in a table. Thus a criterion -referenced test is valid and reliable.

Characteristics of Good Tests

Tests are instruments by which to measure performance. Their characteristics determine whether they perform the function for which they were created. To be effective and perform their educational function, tests must be both reliable and valid.

Reliability

Reliability means that a test must consistently produce the same results under different conditions. A coefficient of reliability may be obtained by correlating two sets of scores, called a *coefficient of reliability.*

Validity

Validity indicates that a test measures what it was designed to measure. Validity relates the outcomes of a particular test to another established test called a *criterion.* The correlation coefficient, which measures the relationship between two tests, is called the *coefficient of validity*. There are several types of validity. *Empirical validity* occurs when the scores obtained on one test are correlated with the scores of the same group of subjects on another available test called the criterion. It is different from *construct validity* in which the test is correlated with a combination of other tests. *Concurrent validity* occurs when the tests are administered at the same time, and the criterion of validity calculates the relationship (see Table 15-1). *Predictive validity* of a test occurs when the test is related to another test (the criterion) administered some time in the future. The test is able to predict future performance on the criterion. For example, scores on high school mathematics tests may predict performance in college mathematics or an IQ test may predict performance in college.

Content Validity refers validity based on the content, such as that of a course, which is not directly measurable. In content validity, the test items must be representative of the entire course. This is done by preparing a table of specifications that outlines the selected content objectives. If a test is valid it will also be reliable (see Table 15-2). In the example of correlation of two tests, Test 1 will be X and Test 2 will be Y. Students will be numbered 1 to 10; the scores of each student will be recorded.

Table 15-1. Tests Demonstrating Predictive Validity Using Pearson's Product Moment Correlation between Intelligence Quotient (IQ) and Tests of Mathematics (Y)

Student	X IQ	Y Mathematics
1	150	90
2	140	90
3	130	80
4	120	80
5	120	80
6	100	80
7	100	70
8	80	70
9	100	70
10	80	70

Correlation may be found by the formula

$$R = \frac{n(\sum XY)-(\sum X)(\sum Y)}{\sqrt{[n(\sum X^2)-(\sum X)^2]\ [n(\sum Y)-(Y)^2]}}$$

Using Excel, the result is:

	IQ	Mathematics
IQ	1	
Math	0.905945	1

Interpretation

The correlation coefficient is 0.90. Hence IQ predicts performance on mathematics tests with ninety percent accuracy.

Table 15-2. Concurrent Validity: Pearson's Product Moment Coefficient Between Two Tests

		X	*Y*
	Student	Test 1	Test 2
A	B	C	D
1	1	90	80
2	2	60	70
3	3	40	90
4	4	80	70
5	5	70	60
6	6	90	90
7	7	70	80
8	8	60	70
9	9	80	80
10	10	70	70

Correlation may be done by using Excel.
Enter data in two columns (*X* and *Y*), and then highlight.
Select *Tool* from the toolbar, then *Data Analysis*. A dialog box appears.
Select *Correlation*. Then input the range C1: D10. Check grouped data.
Enter the input range in B 12. The output appears thus:

	TEST 1	TEST
2		
TEST 1	1	
TEST 2	.030189	1

Interpretation: The correlation (R) between two tests is 0.03.

Measurement and Description of Data

A test is an instrument of measurement. It measures performance of students so as to infers their level of knowledge. Like an instrument, a test is a device that has units of measurement. The unit of measurement on a ruler is an inch; the basic measurement on a test is a *raw score*, which is a unit of measurement. It is a point on a scale such as a test, for example, 90.

Frequency distribution is the process of putting order into an array of scores. It is a systematic procedure for arranging individual scores from least to most in relation to some quantifiable characteristic such as intelligence or scores on a test of mathematics. If the scores are arranged in columns, the teacher may better see how many students received each grade, for example, five students each received a grade of 90 or that frequency of 90 is 5.

Graphic Representation of Data

Test scores may be represented graphically by a line graph or a histogram. A graph visually explains relationships among scores.

Percentile

A person's position in a group with reference to a particular trait such as a score can be considered by percentile. A percentile points out that the person has more or less of a trait than others. Thus a student may be at the eightieth percentile rank, means that eighty percent of the group did less well that the individual. The percentile represents the percent of scores that a particular score exceeds. It is computed by counting the number of scores that the score in question exceeds and then dividing this number by the total number of scores in the distribution, and then by multiplying by 100:

$$\text{Percentile} = \frac{\text{Number of scores below targeted score}}{\text{Total number of scores in the distribution}} = \frac{6}{10} \times 100$$

There are 10 scores in the distribution, 90 is higher than 6 scores. Divide 6 by 10 and multiply by 100. Therefore 90 is at the sixtieth percentile. A score should not be interpreted as on absolute basis; it should be interpreted in relation to other scores in the distribution.

Measures of Central Tendency

Central tendency describes the typical point of a scale. It is the average. A *mode* is the most frequently occurring score. A *median* is the midpoint that divides the distribution into two parts and helps to rank individuals.

The *mean* is the most frequently used measure of central tendency in education. It is a typical score, found by adding all the scores and dividing by the total number of scores. The mean is tied to the positions of the scores and is dependent on each observation. Thus it is the best indicator of a group's performance unless the distribution is skewed. A teacher finds the mean of the test and evaluates the performance of other students whose scores fell above or below the mean. The teacher is better able to evaluate the students' performance and take remedial action. This type of score reporting also facilitates comparisons of the performance of two groups of students. The mean may be calculated by the formula

$$\mu = \frac{\text{Sum of the Values of X}}{\text{Number of Observations}} = \frac{\sum X}{N}$$

The mean of ten scores, 10, 11, 12 13, 15, 16 17, 18, 19, 20 is 15.1.

Variability

The mean tells the typical scores or the central point around which scores center. Individual observations, however, may be widely distributed. It is helpful to know the how widely the scores are distributed around the mean. Measures of variability are the range, which is the difference between the highest and lowest point, and the quartile, which is quarter point and the standard deviation. The most frequently used measure of variability is the standard deviation. *Deviation* is the distance of a score from the mean (μ). Because scores may be negative, the deviations are squared and averaged to obtain a *variance* (σ^2).

The *standard deviation* (σ) is the square root of the variance and is computed by the formula

$$\sigma = \frac{(X - \mu)^2}{N}$$

where σ^2 = Variance,
X = Sum of Observations,
μ = the Population mean, and
N = Number of Observations.

For example:

	Observations		
	X	$X - \mu$	$(X - \mu)^2$
	7	0	0
	6	-1	1
	10	3	9
	5	-2	4
	10	3	9
	4	-3	9
Σ N	42		32
μ	7		

Variance (σ^2) $= \frac{(X - \mu)^2}{N} = \frac{32}{6}$

σ $= \frac{32}{6}$

$= 2.3$

The Normal Distribution Curve

The normal distribution is a family of curves that measures the occurrence of a score on a probability scale. Together they form the normal distribution curve. On a scale each score has deviations that are positive or negative. The mean is used as the starting point, and the deviations as units of measurement above or below the mean on the normal curve.

The curve may be used to estimate solutions to problems. For example, the mean IQ of an eighth grade class is 100 and the standard deviation is 15 IQ points. The question is, "What portion of the class will have an IQ below 114?" The answer may be found by applying the normal curve or bell-shaped curve below.

It is known that below +1 the scores are 34.14%, 34.14%, 13.6%, 2.1%, and 0.1, for a total of 84.08%. Therefore 84% of eighth graders will have IQs below 114.

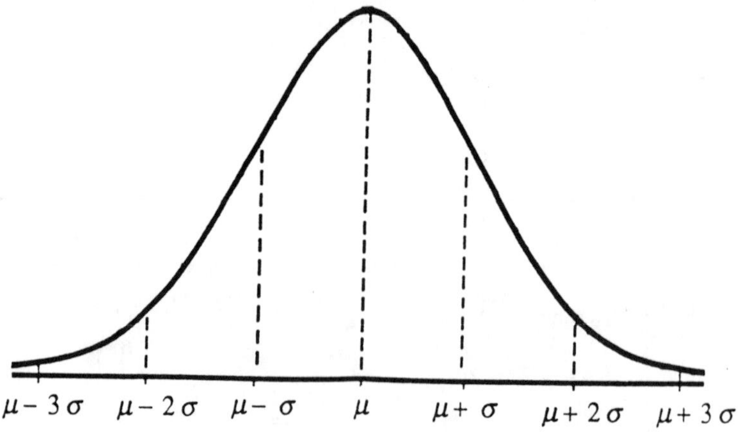

Figure 15-1. The Bell-shaped Curve

Construction of a Multiple Choice Test

Two variables are involved in the construction of a test: the selection of items from content and cognitive processes.

Tests are designed to ensure that certain standards are met with the tester relatively uninterested in the performance in relating individual performance to another. These are criterion-referenced tests. A comprehensive sampling of the instructional objectives is the basis of the tests. The entire spectrum of objectives is examined and selection is made from each objective. In this the tester ensures that every objective is measured.

Categories of Cognitive Process

Just as test items must be selected by content, they must also be selected by cognitive process. A selection by cognitive process must also be comprehensive so as to cover the full spectrum of mental functioning. The idea of classifying mental functioning was discussed at a meeting of the Psychological Association in Boston in 1948. The Association expressed interest in a theoretical framework that could be used to facilitate communication among examiners and proposed that such a framework would promote the exchange of test materials and ideas about testing as well as stimulating research the relationship between examinations and education. The outcome was a system that classified the objectives of education. Benjamin S. Bloom edited the taxonomy of educational objectives, based on cognitive domain. Bloom postulated his taxonomy of cognitive processes, which ranged from the lowest to the highest levels of mental functioning. The classification described in hierarchical form is provided below.

1. *Knowledge* encompasses those behaviors that emphasize remembering or recalling of facts, ideas, terminology, principles, and theories, including

- Specific facts: dates, events, persons, sources of information;
- Terminology: technical terms, relations, concepts in science;
- Vocabulary: quantitative and mathematics definitions and concepts;
- Trends and sequences: processes, directions, movements;
- Classification categories: sets, divisions, arrangements;
- Criteria proportions;
- Methodology: inquiry, techniques, procedures; and
- Principles and generalizations: schemes, patterns, and theories.

2. *Comprehension* is the largest case of intellectual abilities described by Bloom. It involves understanding the material. Bloom identified three types of comprehension: translation, interpretation, and extrapolation. *Translation* means examining symbolic forms such as maps, tables, diagrams, graphs, formulas, and architectural plans and

making decisions about them. *Interpretation* means first translation of symbolic forms, associating two ideas and drawing inferences from them. *Extrapolating* means associating at least two ideas and predicting trends.

3. *Application.* The taxonomy is hierarchical and therefore each category involves skills used in the one below. Thus application requires knowledge and comprehension. It goes further in that it requires the student to know the abstraction well enough to be able to demonstrate its use. This involves applying knowledge to other situations, such as application of concepts, theorems, and principles to a new situation, such as applying laws of trigonometry to the solve practical problems and mathematical principles to new problem-solving process.

4. *Analysis*, by definition, means to break into component parts and examining relationships. Analysis involves relationships, such as facts from hypotheses, conclusions from statements, judgments, cause and effect, logical arguments, fallacies, and assumptions. It also involves analysis of principles in mathematics.

5. *Synthesis.* Just as analysis involves breaking knowledge into constituent parts, synthesis means putting those parts together into a whole so as to create new relationships. The elements are combined into a new pattern, structure, or meaning. It involves

- relating concepts, processes, and relations;
- integrating the outcomes of an investigation into a new whole;
- generalizing relationships into new meanings; and
- production of new meaning from operations.

6. *Evaluation* involves making judgments. It involves examining the skills that had been performed at a lower level of operation and developing an opinion of the value of the material, process, idea, solution, or method. Evaluation uses standards on which to base judgments.

Selection of Items

The emphasis in selecting items for criterion-referenced tests is almost entirely on a thorough sampling of the relevant content and process categories, that is, categories along both dimensions require adequate representation. The method involves using a table of specifications that is based on both content and process dimensions.

The teacher decides the total number of items that should constitute the test. The items are distributed over the matrix, with the greatest number being allocated to knowledge and comprehension and the fewest to synthesis and evaluation. Likewise the largest number of items must be on the easier topics in the course. Table 15-3 provides an outline of such a table based on the unit plan. As shown in the table, there will be twenty-six questions on the test, and they will be allocated to each area of content at each level of mental functioning. The tester will use this information to create multiple choice test, which is most commonly used.

Table 15-3. Table of Specifications

Content►	Rates	Coordinates	Graphing	Slope	Slope/ Intercept	Total
Process▼						
Evaluation	1	0	1	0	1	3
Synthesis	1	1	1	1	0	4
Analysis	1	1	2	1	0	5
Application	3	2	2	2	1	10
Comprehension	3	2	3	3	1	12
Knowledge	5	4	3	3	1	16
Total	14	10	12	10	4	50

Multiple Choice Items

A multiple choice item has a stem, which is a statement, and four options, including one correct answer, one plausible answer, and two distracters. The distracters are incorrect responses that are inappropriate. One distracter shows only partial knowledge or skill of the concept under consideration. The correct answer and the plausible option are the ones that the tester must consider. Examples based on the unit test are listed below.

Knowledge
Read the graph and answer the questions (2, 3)

1. The coordinates of the graph are:
   a. (2, 3)
   b. (-2, 3)
   c. (3, 2)
   d. (2, -3)

2. How many points are required to determine a line graph?
   a. 1
   b. 2
   c. 3
   d. 4

Comprehension
3. On a graph (2, 4) is:
   a. 2 units to the left of the Y axis and 4 units below the X axis.
   b. 2 units to the right of the Y axis and 4 units above the X axis
   c. 4 units to the left of the Y axis and 4 units below the Y axis
   d. 4 units to the left of the Y axis and 2 units below the X axis

Application
4. Mrs James, a real estate agent, makes $150 per week plus a one percent sales commission. If every week she sells one house for $80,000, what will her weekly salary be in dollars?
   a. 8,000
   b. 1,000
   c. 950
   d. 800

5. Anna was assigned to climb a rock of 700 feet. At 1:00 P.M. she was at an altitude of 150 feet, and by 4:00 P.M. she was at an altitude of 300 feet. What was her average rate of change of ascent in feet per hour of the altitude during her climb up the mountain?

   a. 125
   b. 55

c. 50
d. 5

Synthesis
6. The weekly cost of operating a taxi is $50 plus 12 cents per mile. Write an equation expressing total weekly cost, where C is cost and M is miles

a. C= 50 + 0.12M
b. C= 12 + 50M
c. C= 50 + 1M
d. C= 1 + 50M

Evaluation
7. Mr. John James and his wife Anna jog regularly along the same path. John jogs at 5 miles per hour, and Anna at 4 miles per hour. Ana started jogging at noon and John started at 12:30 along the same path. How long was it until the couple met?

a. 2 and a half hours
b. 2 hours
c. 5 hours
d. 1-1/2 hours

8. Based on the information in number 7 above, how far from the starting point will John and Anna be when they meet?

a. 10 miles
b. 20 miles
c. 5 miles
d. 50 miles

ASSIGNMENT

1. Explain the following as they relate to testing and assessment: reliability, validity, intelligence quotient,

2. Explain with examples the following: norm-referenced test, criterion referenced test, objective test, standardized test, power test, speed test.

3. Write definitions with examples for the following: standard deviation, variability, normal curve, and percentile.

4. Choose a unit from the learning standards and write a table of specification involving both content and process factors.

5. Write twenty-five multiple choice items for your table of specifications.

National and Statewide Testing

National Testing

The U.S. Department of Education, through the National Center for Statistics (NCES), conducts the National Assessment of Education Progress (NAEP), which provides insight into the achievement of American students in mathematics and other subjects. In 1993, by Public Law 103-33, Congress authorized the use of the NAEP to evaluate and compare student performance on a state-by-state basis. The tests, which are criterion-referenced and based on the NCTM standards, are reported in what has become known as the Nation's Report Card. The NCES tracks state-by-state scores and the results are reported as a national composite score. It does not give results for school districts.

The results are analyzed in association with a number of variables such as gender, race, income level, home environmental factors or conditions, and attitudes. The population consists of fourth, eighth, and twelfth grade students. The results measure achievement at the basic, proficient, and advanced levels of mathematical functioning.

The National Report Card in Mathematics

The mathematics test is criterion-referenced based on the NCTM standards. The test's mathematics content includes number sense, properties, operations, measurement, geometry and spatial sense, data analysis, statistics and probability, and algebra and functions. The test uses multiple choice test items that require the students to do constructed response.

Sample of Performance Indicators on NAEP Examination: Grade 8, Scale 0 to 500

Below Basic (0–262)

- Find coordinate on number line
- Area of hexagon
- Rounding of decimals
- Use multiplication to solve problems
- Find the area of a figure on a grid
- Identify solutions to linear inequality

Basic (262–299)

- Use nonzero to find length
- Partition area of rectangle
- Identify fractional representation
- Solve problem involving money
- Use pattern to draw path on grid
- Identify acute angles in figures
- Understand sampling techniques
- Solve literal equation
- Graph linear inequality
- Find location on a grid
- Multiply two integers

Proficient (299–323)

- Draw lines of symmetry
- Compute using data from graph of circle
- Reason about the magnitude of numbers
- Read measurement instrument
- Identify functions from table values
- Word problem involving division
- Use scale drawing to find distance
- Determine whether ratios are equal
- Find remainder in division problem
- Find central angle measure
- Find equivalent term in number pattern

Advanced (323–500)

- Compare area of two figures
- List all possible outcomes
- Use scale of drawing to find area

Example of Test Items from the 1996 NAEP Test Examination:

1. N stands for the number of stamps John had. He gave 12 stamps to his sister. Which expression tells how many stamps John now has?

A. N + 12
B. N - 12
C. 12 - N
D. 12 x N

Chapter 16

Alternative Assessment and Reflective Teaching

Alternative assessment, as the term implies, means forms of assessment that are different from the traditional methods, especially multiple choice tests. Whereas traditional tests are objective, based on divergent thinking with one right answer, and involve prescribed content and strict standards, alternative assessment allows for convergent thinking and encourages more creative expressions of individual students. It also allows for evaluation of cognitive knowledge and performance processes.

Alternative assessment has been described in many terms, such as authentic assessment and performance assessment. These terms have a common element and are synonymous. They indicate variants of performance assessments that require students to generate answers rather than to choose among responses that have been preselected for them (Hermann, Aschbacher, and Winters, 1992). Students must accomplish tasks with a measure of complexity and are required to bring previously learned knowledge and skills to bear on a current realistic or authentic problems. Authentic assessment includes demonstrations, investigations, presentations, journals, and complete portfolios.

Eby (1996) suggested that authentic assessment also allows for the teacher to evaluate plans that incorporate learning of content and the thinking process. The task must demonstrate not only what students have learned but also the cognitive process used to complete the realistic task. To evaluate the product, the teacher establishes performance standards that involve comparing performance accomplishments with pre-established learning levels. Eby suggested observation, mastery, and oral reports as methods of authentic assessment.

In the *observation* method, teachers observe students during the process of a lesson to evaluate their level of understanding. Elements considered include participation, attentiveness, or lethargy. Students blank faces or fidgeting during the lesson would indicate a problem.

Mastery of academic content is an exercise in preplanning. The teacher sets the objectives of the lesson, decides on the cognitive processes to demonstrate the performance, and creates an instrument by which to evaluate both performance of content and the cognitive processes.

Oral reports allow the teacher to work with the students to create rubrics for describing the length and format of the reports and the content that must be included. Students have the opportunity to revise their reports before presentation.

Principles of Performance Assessment

1. Link Instruction to Assessment

Every learning unit plan has objectives. The objectives of the lesson should be expressed in criterion-referenced form and the performances defined just as in traditional tests. The difference between the two types of tests is that is traditional tests involve pencil and paper whereas authentic assessment students will be required to create a product to demonstrate the objectives and processes required to produce it.

2. Realistic Product

The main purpose behind authentic assessment is to create real-life performances. The product therefore must be realistic, such as a financial statement describing an arrangement for leasing a car, calculation of the traveling distance from New York to Frankfurt in radians, and output of a computer program used to give directions by the controllers at the airport.

3. Problem-solving Approach

The product must use follow a problem-solving approach and demonstrate the use of content from several disciplines, such as mathematics, science, and technology.

4. Validity and Reliability

The preceding chapter discussed matters of validity and reliability. These requirements also apply to performance assessment instruments. The validity requirement depends on the structure of the test. Content validity specifies that the course content be covered, and process evaluation defines the range of cognitive processes necessary to create the product. The creation of a table of specifications, which embodies a comprehensive list of the content and processes as well as performance specifications, ensures that the test is valid. The method of validating a test is based on Bloom's taxonomy of educational objectives discussed in the previous chapter.

5. Evaluation Should Include Both Outcome and Process

The evaluation should indicate the extent to which the outcomes meet the objectives. Objectives should be clearly defined, standards set, and measuring instruments created. This organization makes the cognitive knowledge easy to measure. Definition of the processes also should make the product measurable.

Methods of Performance Assessment

Having decided on the objectives, the cognitive processes, and the realistic task, the teacher should then create a rubric with which to evaluate the product. The method used should be similar to that used in traditional testing with the main difference being that a product will be the outcome, rather than multiple choice answers.

Step 1. Specify Instructional Objectives

The teacher creates a table of specifications that outlines the content to be mastered upon completion of the task. These will be the instructional objectives of the unit to be assessed. The objectives will be written on the horizontal side of the table.

Step 2. Specify Cognitive Skills Specific to Creating and Presenting the Task

The cognitive processes specific to the task will be written on the vertical side of the table of specifications. It should include skills necessary for completing the task and for presentation. A sample of such a rubric is below. It is for a unit on exponents. The product to be created is described in the final paragraph.

Step 3. Describe Performance Specifications

To evaluate a product requires the use of objective measurements. This is done by specifying the measurement for each level of attainment. The table should embody performance on the cognitive side. That is, the product should embody all the content required to meet the objectives. It should also specify the performance dimension, similar to the Blooms' taxonomy of cognitive objectives. Examples of performance specifications are as follows:

Level 5

- Students have included 100% of the content in the unit and applied them to the task
- The product was complete and the mathematical solutions were ninety-five to one hundred percent accurate.
- The student used higher level cognitive skills like analysis synthesis, and evaluation.
- The examples were relevant and used technology.
- The presentation used graphs, tables, charts, etc.

Level 4

- Students included at least ninety-five of the content and applied them to the task.
- The product was complete and the mathematical solutions were ninety to ninety-five percent accurate.
- Higher-level cognitive skills, such as analysis, were used.
- The examples were relevant and used technology.
- The presentation used graphs, charts, tables, etc.

Level 3

- Students included at least ninety percent of the content.
- The mathematical solutions were eighty to ninety percent accurate.
- Middle-level cognitive skills, such as analysis and application, were used.
- The examples were relevant and used technology
- The presentation used graphs, tables, and charts.

Level 2

- Students included at least eighty percent of the content required.
- The mathematical solutions were seventy to eighty percent accurate.
- Only lower-level cognitive skills were employed.
- Technology was applied.
- The presentation used charts, graphs, and tables.

Level 1

- Assignment was generally incomplete and inaccurate.

Step 4. Decide on Rating Scale

In the end, the teacher has to give the student a grade. The product therefore has to be objectively measured on the content, process, and performance dimensions. A rating scale may be used effectively to measure on all three dimensions simultaneously. As a measure of performance the scale must award scores for each level of performance. These are usually ratings of 5–1 with 5 being the highest. Based on this, above the measurement would be as follows:

Level 5 = 5
Level 4 = 4
Level 3 = 3
Level 2 = 2
Level 1 = 1

Step 5. Create a Rubric and Evaluate the Product

A usable rubric may be written in three dimensions.
X_1 to X_4 indicate the instructional objectives of the content;
Y_1 to Y_4 indicate the cognitive skills employed in creating the product.
Rates 5–1 indicate the score on the rating scale. Thus:

Content	X_1	X_2	X_3	X_4	X_5	
Process						
Y_1	5	5	5	5	5	
Y_2	3	5	5	3	4	
Y_3	5	5	5	5	5	
Y_4	4	4	4	4	4	
Average	4.25	4.75	4.75	4.25	4.50	Grade B+

Examples of Products for Evaluation

Product 1

Create a retirement portfolio consisting of six types of investment instruments. Report the output after forty years. Demonstrate the rules of exponents.

1. *Content dimension* (X)
Content such as (X_1) Rules of Exponent, (X_2) Negative Expressions, (X_3) Graphing of Functions, and (X4) Exponential Equations

2. *Cognitive Processes* (Y)
Cognitive processes to include (Y_1) accuracy of calculations, (Y_2) inclusion of several mathematical rules, and (Y_3) quality of presentation

3. *Performance Specifications*
Five levels of performance (5–1).

Product 2

Make a model of the journey of Apollo 11 from the earth to the moon. Calculate the mass of the earth. Include the speed of the rocket and the influence of gravity. Use scientific notation.

1. *Content Dimension* (X)
(X_1) Rewriting of scientific notation in decimal form; (X_2) computing with scientific notation, using a calculator; (X_3) problem solving

2. *Process Dimension* (Y)
(Y_1) Organizing data, (Y_2) drawing model to scale, (Y_3) using problem-solving techniques, (Y_4) quality of presentation

3. *Performance Specifications*
Five levels of performance (5–1)

Portfolio Assessment

The assessment of a portfolio is a special case of performance assessment. As Herman, Aschbacher, and Winters (1992) suggested, portfolios are collections of student work that are reviewed against criteria to judge an individual student or program. Several aspects are involved in the process:

1. The portfolio includes an aggregation of products that were created over a period of time.

2. Each product was produced based on content, process, and performance specification.

3. The product was reviewed and evaluated by teacher and students and adjustments made. Assessment is therefore process or formative.

4. The final portfolio consists of a selection of the best products placed in a folder. It is a collection of work. Evaluation is summative.

Sample

Table 16-1 below outlines the design of an assessment instrument. It assumes that a number of products will be made, one for each unit of instruction. Content means the instructional objectives which should be specified. Process is should specify the cognitive processes needed to complete the project, and performance should specify the accuracy of the solutions, and the presentation with use of data. Each of these areas should be rated on a scale of 5–1. The average score will indicates the grade.

The assessment instrument combines the comprehensive list of the content, process, and performance specification standards, and is therefore valid and reliable.

Table 16-1. Sample Instrument for Portfolio Assessment

Content	Unit 1(Product 1)	Unit 2(Product 2)	Unit 3(Product 3)
Rating	5–1	5—1	5–1
Entry 1	___content	___content	___content
	___process	___process	___process
	___performance	___performance	___performance
Entry 2	___content	___content	___content
	___process	___process	___process
	___performance	___performance	___performance
Entry 3	___content	___content	___content
	___process	___process	___process
	___performance	___performance	___performance
Summative Evaluation	___	___	___
Average Grade___________________			
Final Grade___________________			

Reflective Teaching

Those who aspire to be teachers like to think that anyone with sufficient knowledge of the subject matter content is able to teach. They believe that teachers are born. Others perceive of teaching as a craft that can be learned with sufficient practice. Professionals, however, regard teaching both as a science and an art. In fact, teaching may be described both as a craft, an art, and a science.

Teaching as a Craft

Alan Tom (1984) described teaching as a craft that includes both mechanical skill and analytic knowledge. Craft creates objects of beauty that bring pleasure and serve a useful function. Craft evolved from the medieval guild system in which new members were recruited as apprentices to learn some skill. It is skill-oriented. Apprentices learn the skills but do not have the knowledge to design and complete something of beauty on their own. Thus if the teacher, who is a craftsman, knows how to apply some skill but does not have the required knowledge and or know all the cognitive ways in which certain things are done. Then, when there is a problem, the teacher cannot take corrective action.

Teaching as an Art

An art object is something of beauty created by an artist who uses either technical or more "free-flowing" processes (Bolin, 1995). The technocrat issues craft and knows how to apply it but does not know how to design. He knows the steps in performance but may or may not produce a good lesson. The free-flowing spirit has an "idea of theme in mind and activities and events emerge" (Bolin, 1995, p. 27). There are no performance standards, however, and the teacher cannot evaluate the lesson.

Teaching as a Science

A scientist knows how to use the technical skills of the subject and how to apply them to the solution of problems. Advances in science have encouraged educators to apply science to the teaching profession. Such terms as *efficiency* and *effectiveness* began to be used in the evaluation of teaching. This means that standards were established and used to evaluate teaching. Educators such as Rice (1893), the journalist, criticized the methodologies practiced in the school system of the age as dehumanizing, dull, and based on memorization and drill. His report was the catalyst for the progressive movement. He also had ideas as to how education should be reorganized. Rice suggested that the best way to produce teachers was to establish practice standards that could be learned by all teachers.

During the progressive era, Franklin Bobbit (1918) introduced ideas for organizing jobs. His activity analysis model led to the practice of specialization in industry. From industry, schools copied methods of organizing the school and the curriculum and the idea of specialization. At that time teaching began mimic science in the area of subject specialization, with each content area having experts. Bolin and Panaritis (1992) suggested that teaching as an instructional problem could be considered quite apart from what was being taught.

Teaching as Art and Science

Educators such as Dewey (1904) had ideas of teaching that were different from the traditional methods that had been criticized by Rice. Dewey suggested that teaching as a science or art were not in opposition to each other. Dewey described teaching as

> . . . an activity includes science within itself. In its very process it sets more problems to be further studied, which then react into the education process to change it still further, and thus demand more thought, more science, in everlasting sequence. Those who master teaching as an art are usually those knowledgeable in their subject matter, while those the craftsmen are those skilled in pedagogy. (p. 235)

Dewey (1904) suggested that

> Those who master technique may appear to have an advantage in their first year of two of teaching, but they will not be as effective in the long run as teachers who learn how to think about their work and hold an experimental attitude. The experimental attitude requires reflection, and making changes as required by the prevailing circumstances. (p. 238)

In essence both the subject matter and pedagogy are important, and both the art and science of teaching are prerequisites of effective teaching. Reflection is key to promoting effectiveness in the teaching process.

The Reflective Process

Henderson (2001) suggested that teachers, like those in the purely creative professions such as writers, painters, and composers, must work to develop their craft. Program designing, lesson planning, and classroom management skills should be continually refined. Reflection is a critical part of the process of teaching and that teachers must engage in a recurring cycle instructional study, application, observation, and reflection. Many young teachers become disgruntled when they are asked to write reflection about their lesson.

The teacher has to be able to answer the question. What happens if you complete a lesson, then proceed to the next lesson without evaluating whether or not it was effective? The purpose of reflection is to help teachers become effective problem-solvers. The process is to evaluate every lesson, see whether the objectives have been attained, and then take corrective action.

Reflection is a type of process evaluation, which is ongoing and creates a cycle of teaching and reteaching. The lesson plan specifies the instructional objective. Effectiveness of a lesson is determined by attainment of instructional objectives. The teacher decides whether or not the objective has been attained and reviews the reasons why. The teacher must examine every component of the teaching-learning process, including

1. The accountability or difficulty of the content to the students,
2. The motivation of the students, and
3. The teacher's own teaching methodology.

Examination of these components must be done in the context of the teacher's knowledge and philosophy of teaching.

The next step is might be one of two directions. If the objectives have been met then the teacher continues with this approach. If the outcome is unsatisfactory, and then the teacher takes into consideration the weaknesses in the lesson and reteaches.

The teacher then reviews the objective or takes other remedial action. This is a continuous cycle.

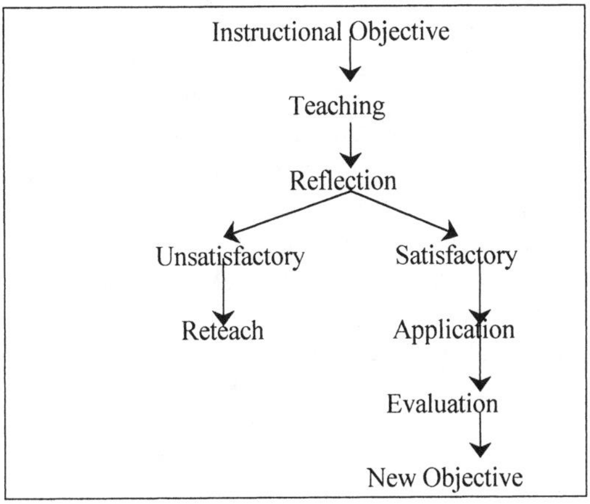

Figure 16-1. The Reflective Teaching Model

Memories are short lived if they are not reinforced and acted on. The teacher's role is to make notes of the main points of the teaching-learning encounter that come to light during reflection. The teacher should make notes in a journal that will be useful for future reference.

Formative or Process Evaluation

Evaluation may be either formative or process and summative or product. *Formative evaluation* is assessment of process. It allows the teacher to determine the level of previous knowledge and the extent to which intermediary objectives are being met. It allows the teacher to make judgments about the effectiveness of resources and the depth of knowledge before moving on. In any process, assessment must take place en route before the process ends. Formative assessment may be quizzes, presentations, or projects. It may also be self-assessment by the teacher, who may change strategies as a result. The teacher performs the process of reflection. At each stage, and whenever it is determined that there is a problem, the teacher stops and reteaches or gives enrichment exercises. The teacher might even have to go back to an earlier stage and give students opportunities to master the foundation principles.

Formative evaluation determines the difficulty of the content, whether weaknesses exist in the students' knowledge or level of achievement, how they react to the teaching process, whether an activity is useful, whether students are able to apply knowledge, or whether the pace of imparting knowledge is appropriate. By practicing reflection and process evaluation, the teacher may initiate corrective action in course content, activities, and the pace of teaching.

As the term implies, *summative evaluation* occurs at the end of the program. Its goal is to determine how well students have attained all the objectives of the course. It also determines whether students have acquired all the knowledge in the plan. Summative evaluation is done using tests, projects, and portfolios. It evaluates efficiency and effectiveness of the school program.

ASSIGNMENT

1. Discuss the terminologies and meaning of alternative assessment.

2. Choose a topic from the school program, and explain how you would use performance evaluation to assess it.

3. Create a rubric for evaluating a portfolio.

4. Create a model of reflective teaching

5. Discuss the difference between formative and summative evaluation

Comparison of Traditional and Alternative Assessment

Traditional methods of assessment are based on subject-centered curriculum and traditional methods of teaching involving authoritarian philosophy and behaviorist psychology. On the other hand, alternative assessment is based on student-centered, standards-based curriculum and problem-solving-discovery methods of teaching.

Traditional Assessment

- Multiple choice tests are the time-honored method of testing. They are objective. Questions are based on a range of cognitive skills from knowledge to evaluation. Questions require one right answer. Thinking is convergent. Traditional tests may also involve short or long responses such as essays based on prescribe knowledge.

- Knowledge is based on the cultural heritage, and facts are derived from the past. Tests may be biased in favor of culture or social class.

- They may be group, or individual tests.

Alternative Assessment

- Alternative assessment may take many forms such as journals, oral presentation, portfolio, or projects.

- They are coherent and require the application of critical, creative, and problem-solving cognitive skills.

- They are interdisciplinary, based on real-life problems or situations that require the convergence of cognitive disciplines to solve.

- They may require cooperation which is a necessary social skill.

Chapter 17

Classroom Research

Educators need knowledge to make decisions and solve problems that arise from classroom processes. This knowledge is most efficient and objective when it is derived from classroom research. Practicing teachers also need knowledge to deal with the day-to-day realities of the classroom, to select the best methodology for instruction, to identify the reasons that some children are not performing well, and to solve behavior problems. Whatever the reason, classroom teachers need information, and it is only those teachers who are in possession of relevant knowledge who can attain the status of consummate professionals. Classroom research begins with knowledge of the scientific method.

The scientific method had its genesis in the European Enlightenment of the eighteenth century. Progress in society and education began when scientific thought emerged as an outcome of the many scientists who recognized problems as an obstacle to progress and created solutions that could become catalysts to change. Enlightenment thinkers rejected the medieval scholasticism that retarded progress and created a new vision of scientific thinking, from the observation and inductive methods of Francis Bacon and the observation and experimentation of Galileo. In education, scientific thinking began when science touched philosophy and gave birth to psychology. At that time, a plethora of research based on scientific method began in education. Education borrowed research methodology from science and applied it to classroom problems.

The average public school classroom is characterized by problems, defined as obstacles to learning. Obstacles in the path of effective student learning are problems to which solutions must be found in the classroom. Classroom research is the methodology that a teacher may use as a tool in the perennial quest for new and improved ways to teach. The process is a search for truth, not infallible truth, but rational truth that provides new ways of examining problems and enlightened ideas and that enable teachers to make decisions. The study of research methods is therefore essential to the teacher who desires to create knowledge, promote progress, and make effective decisions about pedagogy in the interest of the students.

Scientific Research and Education

The purpose of research is to provide knowledge to apply to solving problems and making decisions about education. Research is derived either from experiences or from theory. It begins with questions about a phenomenon. The questions may be empirical, based on experiences, or conceptual and theoretical, derived from literature.

The scientific method applied to education, however, has limitations. Education is regarded as a social science, not a natural science in which methods of investigation are objective and verifiable. The generalizations and theories created from educational research are not without controversy. They are not facts capable of explaining phenomena from which reliable predictions can be made. Further, subjects interact with each other during the research process, making the situation more complex and unpredictable.

Observation in education is more difficult than in the natural sciences. Education deals with human participants who are variable, which makes it impossible to establish laws. Ary, Jacobs, and Razavieh (1972) claimed that the matter for observation is man's response to the behavior of others, which is subjective and expresses motives, attitudes, and values. Further, observers may project their own values to the subject or the researcher may otherwise influence the observation.

Moreover, observable behaviors are difficult to replicate; a human reaction is not as predictable as that of a chemical in a test tube. Hence the findings in research in education are limited to certain samples. Moreover, the interaction between the research and a human participant may create what has been called the *Hawthorn Effect* in which subjects may behave in a certain way just because they know they are being observed. Thus the results are unreliable. The broad area of research must be narrowed down to a highly specific research problem, which must be well defined, and the scientific method must be strictly applied.

Foundations of Research

Scientific research begins with a problem. A problem is any obstacle that gets in the way of learning. Thus all of the factors relating to students, such as behavior, intelligence, academic performance, and social class, that impact on achievement may be sources of problems. Problems may arise from the personal factors of the teacher, such as education, the applied methodology, and the teaching-learning process. Problems may also arise from vicissitudes in the social environment. In fact the sources of problems are myriad. A research problem may be selected from either theory or from the experiences of the researcher.

Theoretical Research

Theoretical research seeks answers that have to be tested. They ask why a phenomenon occurs and usually refer to causes. A theory explains observations. Generally a theory is a generalization that begins with a hypothesis, examines examples using deductive logic, and ends with explanations that create hypotheses from which generalizations may be made. Hypotheses are summary statements that help to guide future research.

This book has discussed several theories of learning that are the result of systematic investigations. A theory gives explanations for relationships and is able to predict future behaviors; theories help in the development of other theories. They create problems that form the basis of research and provide arguments for hypotheses to guide investigation. A good theory is essential to scientific research in education. It explains observable behavior and helps guide the investigation of relationships. It provides the basis for verifying the results and stimulates further discoveries. A theory may therefore be the beginning and end of every research project.

Empirical Research

Research may begin with experiences of the educator, including those of the individual teacher and others from the profession. Experience, as the concept implies, asks questions about experience of a phenomenon. These can be answered by direct observation, which has been previously conceptually defined. Observation of student and enumeration of the instances of the phenomenon provide the answers. For example, consider the question, "How many children are disruptive?" *Disruptive* has been defined conceptually as "shouting out" the answers to questions. The answer to this research question may be found by enumerating the instances of disruptive behavior.

The Research Process

The research process involves the following:

- Identifying the problem;
- Reviewing the literature;
- Constructing a theoretical framework or model, including stating it as a hypothesis or question;
- Creating an appropriate research design;
- Constructing instruments;
- Collecting data;
- Analyzing the data;
- Drawing conclusions and constructing a theory, and
- Reporting the results.

The Research Problem

The teacher needs to apply certain criteria to a problem to ensure that it is worthwhile for investigation. Ary, Jacobs and Razavieh (1972) suggested that judging a problem is usually a matter of individual judgment and is subjective. Notwithstanding, there are certain criteria to be followed.

The answer to the problem should have theoretical or practical implications. The solution should make a contribution to educational thought and practice. The solution should fill a gap in the extant body of knowledge. Educators should be able to examine the findings and implement solutions in a classroom situation. The research should produce generalized knowledge that explains new relationships and solutions that can be implemented.

The problem should not only answer questions but should generate new questions. Other researchers should be able to do follow-up work or replicate the research. The research should contribute to the evolution a good theory.

The problem should be realistic. A good problem must show a relationship between two variables. The variable must be quantitatively measurable. The researcher must thus be able to define the variable operationally.

Review of Related Literature

There is an aphorism that states, "There is nothing new under the sun." This is true as every problem that an individual may encounter has also been experienced by someone else; moreover, some research has been done on the problem or aspects of it. There may be gaps in the body of literature, however, as research may not have covered every topic. It is therefore important for the beginning researcher to look to the body of knowledge on the topic and review the literature to set the framework for the research.

The related literature will prescribe a theoretical framework for conducting the research. The literature describes what research has been done in the field, identifies gaps in knowledge that are researchable, and suggests the theories for conducting new research. The theory suggests new questions for investigation and from which hypotheses may be stated and definitions described. The definitions lend ideas to the type of instruments that may be used to conduct the research. The literature describes instruments that have been used and have proved useful. Hence the related literature is the starting point for all new research.

Most important, the review of the literature should be of actual studies that have been done. Theoretical literature helps to define the framework, but the body of the review should be of previous empirical studies. The researcher should examine journals and other literature in the field. The literature should be arranged in some order. The researcher should select themes based on the variables being considered for study, and these themes should form the basis for the theoretical framework, the research questions, and the hypotheses that will guide the research study. The abstracts of the materials being reviewed will help the researcher determine whether the work is relevant to the current project. The researcher should first investigate the most recent studies in the field and then work backward to earlier research. Notes on cards and careful recording of the bibliographical information will help the research maintain organization and attend to details needed during later phases of the research. To avoid plagiarism, the researcher should be sure to note the sections that are paraphrases and those that are quotations. The complete file of cards should be organized thematically for reference.

The Theoretical Framework

The theoretical framework describes how the variables are interrelated. Persaud and Figueroa (1972) suggested that by specifying which variables are related, the researcher provides a systematic view of the phenomenon. These relationships generate hypotheses to be tested. A practical way to determine whether a problem is researchable is to express it in the form of a question that asks about a relationship between a dependent variable and one independent variable.

Labeling and Defining the Variables

The researcher should begin by identifying the variables to be studied. This means defining the variables operationally to ensure that they are measurable. Tuckman (1972) stated that in educational research many operational definitions are based on character traits such as intelligence and personality. For example, an intelligent student may be defined as having a good memory, the ability to reason logically, a large vocabulary, good mathematical skills, and the ability to engage in spatial reasoning. Intelligence can be defined as a score on an IQ test. An effective teacher may be defined by the observable behaviors of the teacher, such as encourages students, gives praise, calls on all students equally, provides suitable activities, makes learning relevant to life, and asks inductive questions.

Types of variables include dependent, independent, intervening, and control.

The *dependent variable*, also called the response or output variable, is the subject being studied, such as student achievement, attitudes, or behavior. The dependent variable will be measured, observed, and manipulated so as to estimate the effect of the input or independent variables. As the name implies, the dependent variable is a consequence or outcome of action of independent factors. Its value varies, as the independent variables vary. In effect the dependent variable is dependent on input factors.

Independent variables are the input variables or factors that when manipulated affect the dependent variable. They determine the relationship of the independent variable being studied. When manipulated, independent variables create a product. They are the X and the dependent variable is the Y in a functional relationship. The expressed relationship between the independent and dependent variables is the basis of the question, the hypothesis, and the theory that guides the research.

Stating the Questions

Having defined the variables, the researcher then asks questions to define the relationships among the variables. For example,

- What is the relationship between intelligence and academic performance?
- Is there a relationship between student achievement and dropout rates?
- Is there a relationship between socioeconomic class and student achievement?
- Are students who learn by rote able to solve mathematical problems?
- Do students who are taught by the direct instruction perform better than those who are taught by problem-solving methods?

The Hypothesis

The research questions may be expressed in the form of a hypothesis, which is the general form for testing it statistically. A hypothesis is the question in statement form, which expresses the relationship between the variables. From these statements of a relationship between the variables, a generalization may be made and theories developed. For example, student performance is related to the methodology of the teacher. The statement of the hypothesis assumes that predictions may be made based on the relationships among the variables.

The Research Design

Persaud and Figueroa (1972) suggest that a research design is a strategy for observing and measuring the variables under varying the research conditions. It permits the research to examine the interrelationships of the variables and the effects of the independent factors on the dependent variable. The design must permit the examination of the experimental and control variables so as to ensure that the results are reliable and true. Gall, Borg, and Gall (1996) suggested that experimental designs may be classified by single groups, with no control group, and control group designs.

Experimental design is usually the chosen method of classroom researchers. Gall, Borg, and Gall (1996) and Kerlinger (1976) suggested that experimental designs may be classified by single group, which has no control group, and control group designs.

Single-group designs with no control group may take the form of a single-shot group study or a one-group sample over time.

A *one-shot group study* involves an experimental group X and a posttest O. The group interacts with the experiment. The experiment is valid because the group is the same throughout the experiment, being unchanged for history, selection, and mortality. The design is

Experiment	Outcome
X	O

In a one-group pretest/posttest study design, only one group is involved. There is a pre-test O_1, an experiment X and a posttest O_2.The design is:

Pretest	Experiment	Posttest
O_1	X	O_2

As with a single group, the group remains the same throughout the experiment and so there is control for history, maturation. Other factors which affect validity are the interaction of the two tests.

One group sample over time is a longitudinal study that involves administration of several pretests followed by the experiment and then several posttests. The sequence is pretest followed by experiment and then posttest and repeat the process.

Pretest O_1	Experiment X_1	Posttest O_2
O_3	X_2	O_4

Control group designs include posttest-only with control group, pretest-posttest control, and four-group factorial designs.

The *posttest only with control group* design is based on two randomly chosen groups (RS). One group is experimental (X) and the other is the control group. The experimental group is treated, and the control group receives no treatment. The design appears thus:

RS	X	O_1
RS		O_2

The experiment is to test the effects of interactions with the treatment. Hence both groups are randomly selected and no pretest is given to either the experimental or control group. The lack of pretest ensures that the groups are controlled for mortality and randomness and that the groups are equivalent, biases removed, and the experiment is valid. Thus the effect of the treatment may be treatment evaluated and considered reliable. The effect of the experiment is evaluated on the basis of the comparison of output 1 and output 2. The analysis involves a comparison of the means of the two groups.

Pretest-posttest control research designs involve two randomly selected groups (RS) groups, one experimental group (X), and a second control group. The experimental group is treated and the control group is not treated. Both groups are given a pretest and a posttest. The design appears thus:

RS	O_1	X	O_2
RS	O_3		O_4

The control group is similar in every way to the treatment group in every way except for treatment. The similarity ensures that the groups are equivalent, in history, maturation of the participants, and retrogression of behavior, and that the experiment is valid and the outcome is reliable. As in the natural sciences, control is a very important issue. No experiment can be valid without carefully control of all the extraneous factors that might impact on the participants and change the outcome.

Four-group factorial designs are adaptations of the control group designs. There are four groups, (Y_1) and (Y_2). The groups are randomly selected (RS) and each experimental group (X) has a control group.

RS	X	Y_1	O_1
RS		Y_1	O_2
RS	X	Y_2	O_3
RS		Y_2	O_4

This design makes it possible to evaluate the effect of the experiment individually and in combination. This can be done by creating a matrix.

X

	X_1	X_2
Y_1	O_1	O_2
Y_2	O_3	O_4

Y

With the control group, the teacher randomly selects the students for the groups. Each group takes a pretest. The experiment follows, and then a posttest. The researcher compares the means of the two groups and conducts a statistical analysis to determine significance. Mathematics students will have taken a course in statistics and may easily analyze the data.

Analysis of Data

Experimental data may be analyzed by (a) descriptive statistics or (b) inferential statistics.

Descriptive statistics include scales of measurement, representation of data, measures of central tendency, variability, and measures of correlation:

- *Scales of Measurement* include nominal, ordinal, interval and ratio scales.

- *Representation of data* is by frequency distributions and graphical representations, such as histograms and line graphs.

- *Measures of central tendency* include mean, mode, and median.
- *Variability* includes range, quartile deviation, variance, standard deviation, standard scores, and the normal curve.
- *Measures of Correlation* includes Pearson's product moment correlation and completes the measures for describing data.

Inferential statistics refer to the characteristics of a sample drawn from a population. A population is the total universe of the group being studied, such as all seventh graders or the body of university students. A good sample is usually randomly drawn and is a representation of the larger population. Hence from the sample, inferences may be made about the characteristics of the entire population. The laws of probability guide estimates of results. This allows for errors to be made. Statistical tests are therefore performed on the assumption that there may be errors, and one accepts the inference based on the probability that the phenomenon will occur within a level of confidence. Confidence levels are based on the probability curve and are estimated to five, ten, one hundred, or one thousand probability of its occurring. Inferential statistics uses such tests as the T test and analysis of variance. Analysis of multiple factors may be done by using the chi square test and multiple regression analysis.

Testing Hypotheses

Analysis of data begins with testing the hypothesis. There are two types of hypotheses.

The *substantive hypothesis* is a statement that expresses the relationship between the dependent variable and the one or more independent variables. To test the hypothesis that there is a relationship between the variables, the hypothesis has to be rewritten in null form. A *null hypothesis* states that there is no relationship between the variables. Thus stated, the hypothesis can be tested statistically to determine the probability that the hypothesized relationship is supported by fact. Statistics can determine the probability that the relationship differs from zero.

To test a hypothesis, the researcher states the consequences that would result if the relationship is true as hypothesized. For example, a teacher selects two groups of students, an experimental group and a control group. The study is arranged so that the only difference between the groups is the method of instruction. The experimental group is taught by method A and the control group by method B. At the end of the experiment the teacher gives a test and observes that the mean score for students taught by method A is higher than the mean score of students taught by method B. The teacher has to explain the difference in the scores.

The researcher has to have a high level of confidence that the difference was caused by the experimental method rather than as a chance occurrence. This is possible as the other factors such as intelligence could make the difference in performance. Thus the researcher tests the hypothesis statistically by employing a null hypothesis, which states that there is no actual relationship between the teaching method and the students' performance and that any difference was caused by chance.

The *research hypothesis* (H_1) states that students who are taught by method A perform better that those who are taught by method B.

The *null hypothesis* (H_0) states that there is no relationship between the method of teaching and student performance and any relationship is a function of chance, or the mean performance of all seventh graders taught by method A is equal to the mean performance of all seventh graders taught by method B.

Levels of Significance

The null hypothesis is tested statistically for significance to determine the probability that the performance differs from zero, which would indicate that there is no relationship. There is a point at which the null hypothesis must be rejected. The level of significance is the point at which the probability that the difference is a chance occurrence. Confidence levels are tested by the T test for small samples or by the F ratio. Both tests are probability distributions. Generally confidence levels used are .05, .01. or .001 levels of significance.

Example 1: Analyzing Results of Data Pretest and Posttest on Experimental Design

The researcher gave a pretest at the beginning of the semester in January. The mean score for the group was 70. The researcher taught a mathematics unit using the project method and gave a posttest at the end of the semester in June. The mean score on the posttest was 88. In addition to gaining new knowledge, students in both groups had gained five months in age. Therefore the gain on the posttest and the age gain would be about equal to what would be expected by chance.

Chi-square (χ^2) test may be used to compare the frequency observed and the frequency expected.

$$\chi^2 = \frac{\Sigma(fo - fe)^2]}{fe}$$

$$fo = 70 \text{ and } 88 \; fe = 50 \text{ and } 50$$
$$\chi^2 = .66$$

where fo = frequency obtained
fe = frequency expected
χ^2 = Chi-Square

A check of the χ^2 tables shows that this result is significant.

Example 2: Statistical Analysis of Pretest-Posttest Control Group Experiment

The first step is to compute descriptive statistics. The mean scores on the experimental and control groups are almost identical. The experimental group has a mean of M = 70, and the control group has a mean of M = 70.1. The means of the posttest are very different. The experimental group posttest mean was M = 88 and the control group was M = 72. Statistical analysis must be done to test whether these differences are a chance occurrence or are significant, that is different from zero. Because there may be initial differences in the groups, the test to be used is the analysis of covariance.

Raw Scores on Mathematics Test

	EXPERIMENTAL GROUP		CONTROL GROUP	
Student	Pretest	Posttest	Pretest	Posttest
1	81	90	88	85
2	54	91	46	86
3.	69	89	70	80
4	82	90	78	79
5	74	88	79	68
6	60	80	59	48
Sum	420	528	420	446
Mean	70	88	70	74.3

F = 94.38** is $p < .001$, which is significant.

Figure 17-1. Videoconferencing Classroom at Old Westbury

Part 4

Technology in Mathematics

Chapter 18

Technology and Mathematics

Technology had its birth in mathematics, and in the process of its development technology has become the handmaid of mathematics. Technology is used to perform the empirical functions of mathematics that once were done mentally and by observations and instruments like calculators, and compasses. Mathematical principles such as logarithms once performed the function of calculating large numbers, and engineers and scientists used geometrical instruments to attain the levels of precision required in their professions. Computer programs, speedy calculations, and real-life problems employ technology to configure, analyze, and discover solutions.

As discussed in Chapter 1, statistics from the *Condition of Education* (U. S. Department of Education, 2000) indicate that the number of students graduating from colleges and universities declined during the latter half of the twentieth century, the same period in which technology emerged. The percentages decreased from 3.0 percent graduating in 1971, to 1.3 in 1982, and to 1.2 in 1995. Statistics also show that the percentages of students pursuing studies in engineering and computer studies increased significantly during the same period, fluctuating from 6.2 percent in 1971, to 10.5 in 1982, 9.0 1992, and 8.9 in 1995. The inferences may be made that students have diverted their talents from studies in mathematics to more pragmatic pursuits in the practical applications of mathematics. The figures indicate that technology has overtaken mathematics as the foundation subject. In the future there will be less need for personnel with skills in pure mathematicians, and those who do mathematics must be competent in technology, such as computer engineering and programming.

Mathematical form and concepts precipitated the emergence of technology. Mathematical form, according to De Morgan (1972), developed precision and logic in calculating Cartesian geometry, calculus, and analysis that are essential to technology. Charles Babbage and Augustus De Morgan who described the connection between mathematics and logic. Babbage began constructing machines to illustrate this connection between syllogistic reasoning and the Venn diagram. The work of Baggage and de Morgan was taken a step further by George Boole who invented Boolean logic, which has its basis in the realm of algebra. According to De Morgan (1972), the purpose of Boolean algebra is to find unknown quantities by solving equations and making generalizations into a single formula. Boole invented the idea of examples and nonexemplars, represented by X and Y translated into (1- X) and 0, which could be translated into the equation $X(1 - X) = 0$. Boole also introduced the idea of sets and intersection.

De Morgan (1972) stated that between 1830 and 1930 the Germans George Cantor and Richard Dedekind introduced the idea of sets and pairing of integers with their doubles numbers, for example, (1, 2) (2, 4); the idea of infinite set, and the addition of a number line. Cantor suggested the idea of numbers (0, 1), the diagonal, and linear programming. Leopold Kroneckor took the next step in mathematical application by introducing analysis and topology.

The reversion to mathematics logic, derived from Aristotle, furthered the march to technology. In a book called *Principia Mathematica* (1904), Bertrand Russell and Alfred Whitehead developed the idea that mathematics and logic are almost synonymous. They invented symbolic logic, which has become a topic in high school mathematics books.

Bohl (1984) and Struik (1987) outlined the stages of the development of the modern computer. Early efforts to invent a compute began in 1642 with Blaise Pascal who developed an adding machine with rotating teeth that represented the number 0 to 9. The machine had rotating gears, and as the gear rotated to 9, the next digit was the one to the left of the tooth. Gottfried Leibnitz developed the next step by inventing an adding machine that preformed multiplication by repeating addition and subtraction. In 1890 Herman Hollerich created a machine designed to process census data using punch cards. Hollerich's efforts reduced the processing time required for census by two-thirds over that required a decade earlier, reducing it to two and a half years.

Hollerich established the Tabulating Machine Company (CTR) to produce and market punch cards. CTR became International Business Machines, known as IBM, with Thomas J. Watson as its president.

After World War II, a group of Harvard students and IBM worked to develop faster machines for processing data. Through their efforts, the first computer, known as Mark 1, which accepted punched data, was born. At the same time professor John Atanasoff of Iowa University introduced another type of computer based on the binary system and the first computer came into being in 1942. The new computer called the Atanasoff-Berry Computer (ABC), which could solve linear equations.

The next step came when the U.S. army developed the Electronic Numerical Integrator and Calculator (ENIAC), which used electromagnetic relays and was very swift. A significant development came when the Electronic Delay Storage Automatic Computer (EDSAC) was completed at Cambridge, England in 1949. This significant development represented an introduction of a complete program with sequential instructions and an internal storage unit or memory. The first commercial computer was installed in the U.S. Bureau of the Census in 1951. By the turn of the century, commercial computers are available at reasonable prices and most schools are equipped with computer laboratories.

The history of mathematics has made us aware that in life there will be need for mathematics-related skills such as problem identification and restructuring. According to Willoughby (1990), there will be a continuing need for skills such as recognizing problems, developing reasonable solutions, interpreting solutions and re-examining and improving them, and considering related problems. In effect the teaching of computer languages began in 1642 with Blaise Pascal, who developed an addition to practical skills of mathematics that enables transformation of real-life problems into computer programs.

Modern computer programs can solve every type of mathematical problem. Computers apply data analysis, financial information, and statistics to the solution of mathematical problems such as calculus and differential equations. A computer program can perform all these functions with the click of an icon. The individual, however, must be able to interpret the solutions and hence students must be trained in mathematics to be able to use these programs effectively.

Some topics in mathematics naturally lend themselves to use by technology. Heutinch and Munshin (2000) suggested that technology makes some traditional topics more important or less important and introduces new topics. They stated that technology makes it possible to teach topics in discrete mathematics such as convergence and divergence by watching patterns on the computer. Indeed teachers of mathematics have indicated that it would be impossible to teach calculus without a graphing calculator. Calculators and spreadsheet programs enable understanding of the limit and the area under the curve, which are abstract and can be challenging to comprehend, and to see a function change to a graph. Computers can demystify the properties of three-dimensional figures.

Technology in mathematics education should consist of two aspects: writing programs and operating application programs

Writing Programs

Writing computer programs requires mathematics skills. Indeed writing programs allows one to evaluate the writer's level of mathematical skill. Hence students who are taught to write computer programs are those who know the most mathematics and also those who will learn mathematics in depth. There was a time when students learned Beginners All-purpose Computer Language (BASIC), which has application in business, economics, finance, and the social sciences. in schools. This practice has been largely discontinued, and instead students are taught to use application programs. BASIC might be obsolete as a language, but it still has value as a mathematics program. Moreover C and C++ are programs that are widely used by mathematics majors in colleges and should be studied in high schools.

Operating Application Programs

Most programs that are used in schools are application programs. Bohl (1984) suggested that these programs help students learn several skills. The problem-solving process involves gathering data and making inputs into the computer. The computer solves the problem. For example, there are programs to calculate mortgage interest. The computer may be used for inquiry, that is to find new data. The Internet is an excellent library for finding new information.

Most of all there are programs that simulate real-world data that can be used for practice and for evaluating data by tests. The programs in this chapter are an introduction to using the computer program Excel to solve problems in data analysis, finance, statistics and descriptive statistics, and mathematics. These problems are worked out below with accompanying instruction from which the reader may learn how to use Excel functions to solve problems in mathematics.

Using Internet Resources to Solve Mathematical Problems

Websites on the Internet change from time to time. There are, however, sufficient information and resources available on the Internet for one to use them effectively to solve problems in mathematics. The following lesson plan exemplifies this principle.

1. *Topic*: Calculating Pi
   Grade 7

2. *Objectives*:
   - To help students identify pi in various situations
   - To calculate pi with accuracy
   - To calculate the relationship between pi, circumference, and diameter.

3. *Instructional Strategy*: Guided Discovery

4. *Resources*: Internet resources on blackboard.com

5. *Previous Knowledge*: Students know how to calculate ratios.

6. *Procedures*

   A. *Motivational Problem*
   Show pictures of buildings and windows with Venetian glass that are constructed based on circles. Ask students to consider what dimensions the builder had to know to construct them.

   Students will view video, *The Story of Pi*, which describes the uses and development of pi over several years and pi in various cultures such as Egyptian, Hebrew, Chinese, Japanese, and Greek.

B. *Developmental Activities*
Activity 1: Ask students to identify on a draft of a building on the Internet some of the measurements that the builder would have to know.

Go to www. webmath. com and write the answer on the chart.

Choose one:

Circumference is:	_______(1) Measurement around
	_______(2) Measurement across
Diameter is:	_______(1) Measurement around
	_______(2) Measurement across
Radius is:	_______(1) Half of the diameter
	_______(2) Half of the circumference
Pi:	What is this?

Activity 2: The next task will be to examine some of these measures and find the value of pi.

Follow the sequence on your computer: Start-Accessories-Calculator. Using the calculator, complete the chart.

Container	Circumference	Diameter	Ratio of Circumference/ Diameter
°	7.5	2.3	?
☺	100.00	31.8	?
Ø	393.38	125.24	?

Question: What do you notice about the ratios of the circumstance to the circle?

Answer:_______________________________________
(They are constant.)

Look for the definition of pi at www.ask.dr.math.com

Write the definition here:

____

Activity 3: Students will learn how pi has been used in history. Students will do research on the Internet. The constant ratio has been used by mathematicians over the centuries.

Go to www.webmath.com and answer the questions.

Mathematician Value of pi

____

____

____

What is the current value pi?________________________

Activity 4: Students will look in the environment and identify situations in which pi may be used.

____

____

____

7. *Guided Practice*

a. Write definitions as follows:
C = ____________________(in terms of diameter)
C = ____________________(in terms of radius)

b. Calculate the dimensions.

1. Find the diameter of a circle whose radius measures
a. 4_______________ b. 14___________ c. 4.5 _____________

2. Find the radius of a circle whose diameter is:
a. 4______________ b. 7______________ c. 3.5____________

3. Find the circumference of a circle when the diameter is
a. D = 10, then C =_____________________________________
b. D = 16, then C =_____________________________________

8. *Independent Practice*

Students will do the following problems for homework.

Go to www.webmath.com and do the problems.

9. *Evaluation Quiz*

1. Define pi ______________________________

2. Given radius, write a formula for calculating circumference ______________________________

3. If the radius of a circle is doubled, what change takes in the diameter? ______________________________

Chapter 19

Problem-solving Using Spreadsheets

A. Learning Excel Basics

Entering Numbers

Access a worksheet
Enter the numbers by using TAB and ENTER on the keyboard.

A Grade 7 Mathematics Test:

	A	B
7	Alice	90
8	Ariana	90
9	Brian	86
10	Beverly	84
11	Cecile	80
12	Dawn	70
13	Effie	80
14		580

Using Autosum

Click on B14 to make it the active cell. Click on AUTOSUM (Σ) on the toolbar. The function = SUM (B7:B13) surrounded by a marquee appears. Press ENTER and the sum will appear in B14.

Using Formula to Find Sum

Enter data	B
24	90
25	80
26	90
27	89
28	71
29	420

Make B29 the active cell. In B28 enter the formula =B24+B25+B26+B27+B28. Excel displays SUM on the toolbar and the result appears in B29.

Using the Paste Function to Find Average Mathematics Test Scores

Click on B9 to make it the active cell. Access INSERT on the toolbar. Click on the PASTE function (Fx). A dialog box appears. Select AVERAGE, then OK. In the Numbers box, enter the range B2:B8. Click on OK. The result appears in B9.

Copying a Formula

Enter the data:

	A	B	C
56	X	Y	Average
57	90	90	90.0
58	90	85	87.5
59	86	89	87.5
60	84	80	82.0
61	80	91	85.5
62	70	78	74.0
63	80	78	79.0

Find the average of the scores in the exercise above.

Click on D57. Access INSERT, then the PASTE function (Fx). Click on AVERAGE and then enter the range A57:B63. The average appears in D57. With the cursor in D57, click on the COPY button (right of the scissors icon). Drag from D57:D63. A marqueee appears around the cells to be copied. Click on the PASTE button (right of the COPY button). The formula is pasted and the averages appear.

Applying Style to Data

75	$ 90.50	91%
76	$ 86.30	87%
77	$ 87.30	88%
78	$100.00	100%

Highlight the data. Access the formatting toolbar and click on $ and %, and DECREASE DECIMAL respectively.

To add bold, click on (B) Bold; for underlining, click on (U) underline.

To align data within the cells, click on the left button to left align, the center button for center align, and the right button for right align.

Creating Charts

Enter and highlight the data. Select INSERT from the toolbar. Select CHART, then CHART TYPE, then next, then finish. The chart appears.

100	X	Y
101	90	80
102	80	70
103	70	60
104	60	68
105	50	51
106	41	51

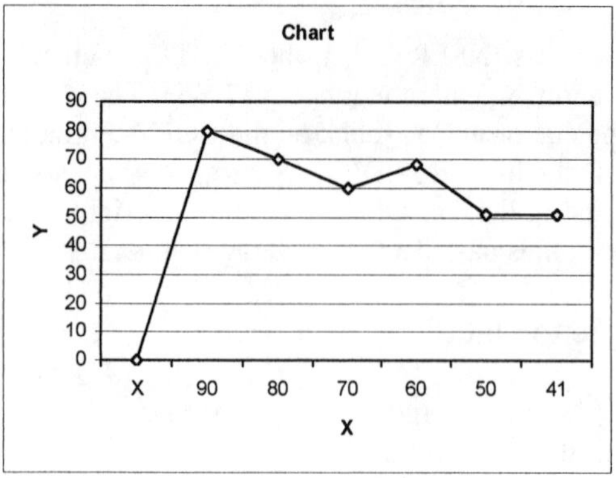

Printing

Highlight the data to be printed. Access FILE from the standard toolbar. Click on the print area. Excel indicates the print area, which is anywhere there is text. Indicate the pages to be printed and then click on PRINT.

B. Algebra

Linear Equation of the Form Y = MX + B

SLOPE (known Ys, known Xs).

Problem: John began to walk up a hill. At 1:00 P.M., he had ascended 10 miles, and by 7:00 P.M., he had completed 55 miles. What is his average rate?

Values of Y (Time)	1	3	5	7
Values of X (Distance)	10	25	40	55

Task 1: Create table with values. In H21, enter the formula =SLOPE(C21:F21, C20:F20). The program returns the value 7.5. In H21, enter the formula =INTERCEPT (C21:C20, F21:F20). The program returns the value 2.5.

B	C	D	E	F	G	H
Table of Values						
Values of Y (Time)	1	3	5	7	Slope	7.5
Values of X (Distance)	10	25	40	55	Intercept	2.5

Task 2: Enter the table of values. Access the chart wizard from the toolbar and select a line graph. Double click on the chart. From the toolbar, access INSERT and then TRENDLINE. From the dialog box, choose the linear graph. Complete the graph and click on OK. Activate any marker on the chart again. Select CHART option. On the series name, click on X and Y and the data points will appear.

1	3	5	7
10	25	40	55

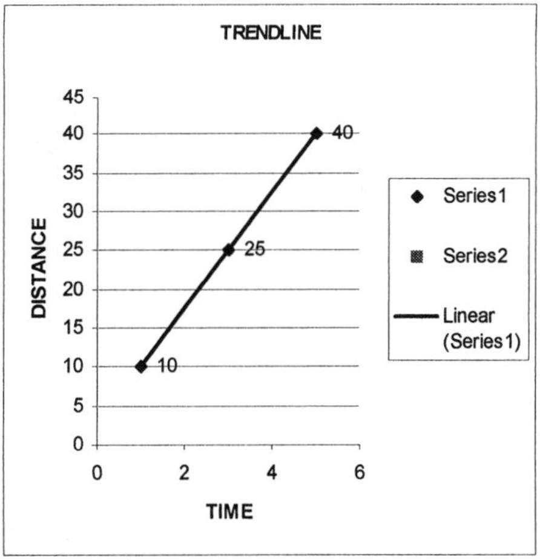

Task 3: Make the trendline regress to zero. Click on any marker on the plot area to make the chart active. Select CHART from the toolbar and choose Type. Click on TRENDLINE/REGRESSION. Click on OPTIONS and label the chart. From the dialog box, select FORECAST. The regression line trends to zero.

Task 4: Display the equation. Enter the X and the Y values on the chart and create a scatter diagram. Select INTERCEPT from the dialog box and enter zero. Click on DISPLAY EQUATION and OK. The equation below will display with values.

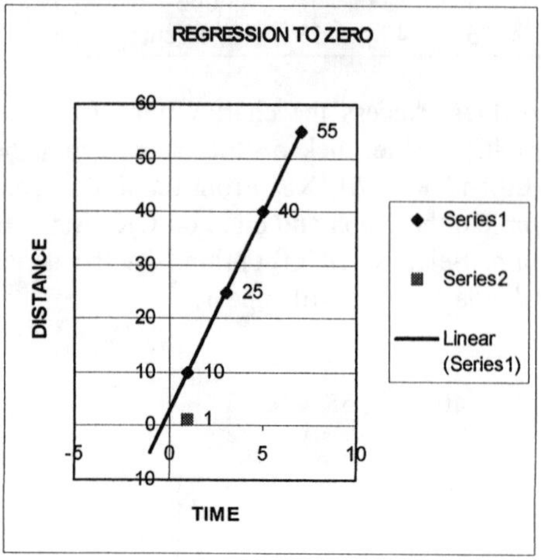

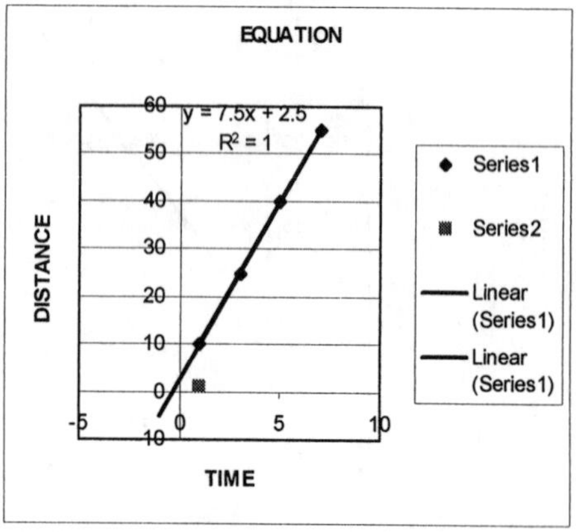

Application of a Linear Equation

This exercise calculates telephone bills for the New Age Telephone Company. The telephone bill is calculated by the formula Y = A + BX or Y = MX + B.

Total cost is represented by	Y	in Column H
Connection cost	A	in Column A
Rate	B	in Column B
Cost for additional minutes	X	in Column C

Connection cost	= 16 cents
Rate per minute	= 5 cents
Additional minutes	X

Enter data:

Column 1	Names of customers	
Column 2	A: Connection cost	A27
Column 3	Rate per minute	B27
Column 4	Additional minutes	C27
Column 5	Total Cost	H27
Column 5	In H27 enter formula =(A27+(B27*C27)/100))	
	In H28 enter formula =(A28+(B28*C28)/100))	

Drag over all numbers.

		A + Initial Cost	(B * Rate per Min.	C) = No. of Minutes	Y Total Cost $
27	Ari	10	5	0	0.10
28	Annie	10	5	50	2.60
29	Barry	10	5	200	10.10
30	Bill	10	5	115	5.85
31	Carl	10	5	125	6.35
32	Clara	10	5	70	3.60
33	Dora	10	5	190	9.60
34	Effie	10	5	100	5.10
35	Gena	10	5	600	30.10
36	Hill	10	5	290	14.60
37	Joy	10	5	98	5.00
38	Kerry	10	5	200	10.10
39	Lana	10	5	500	25.10
	Total				128.20

Solving Simultaneous Equations

Problem: John had $10,000 invested in two certificates of deposit. One CD paid annual interest of 5% and the other paid 5%. The annual statement shows a combined interest of $510. How much of the $10,000 was invested in each CD?

Equations	(two CDs)	X + Y	=	10,000.00
	(two interests)	0.05X + 0.06Y	=	510.00
Rewritten:		X + Y – 10,000	=	0
		0.05X + 0.06Y	=	0

Formulae for equations			
First equation	C26	=	B26+B27-10,000
Second equation	C27	=	.05*B26+.06*B27-510
	E26	=	ABS(C26)
	E27	=	ABS(C27)
	D27	=	E26+E27

Create a table describing the two equations (see below). Enteer X in A26 and Y in A27. In B26 and B27, enter zero to indicate the target values of the equations. Enter fhe formulae for the euqations.
In C26 enter =B26+B27-10,000
In C27 enter =.05*B26+.06*B27-510
In E26 enter =ABS(C26)
In E27 enter =ABS(C27)
In E28 enter =E26+E27

23	A	B	C	D
24		Table for Simultaneous Equations		
25		Variables	Equations	Absolute Values
26	X	0	-10000	10000
27	Y	0	-510	510
28			Sum>	10510

Activate cell E28 (the target cell). From the toolbar, access TOOLS, then SOLVER. In set target cell, enter E28. In value, enter 0. By changing cells enter E26:E27 or click on GUESS. Click on SOLVE and the solution will appear. The results will be similar to those shown below.

23	A	B	C	D
24		Table for Simultaneous Equations		
25		Variables	Equations	Absolute Values
26	X	4675.335767	5.43404E-06	5.43E-06

27	Y	5324.664238	4.565943439	4.56594344
28			Sum>	9.9999434

The amount of $10,000 invested in two CDs is allocated as shown above.

Solving Quadratic Equations

Problem: A department store had a square display with dimensions X * X and area = X^2. After extending the window, the new area was $X^2 + 5X + 6$. What are the new dimensions of the display window?

Objective: To find the roots of the equation $Y = X^2 + 5X + 6$.

SOLVER is the appropriate tool for this problem. Note that SOLVER is an add-in program. To install SOLVER, click on TOOLS, ADD-IN, and complete the dialog box. SOLVER is available for Windows 2000.

First, make an initial guess as to the roots of the function and then use SOLVER to calculate the roots. The aim is to move the roots to zero. First enter one coefficient as zero and guess the other. The guess is 2, as follows:

	A	B	C	D
26		Roots	Equation	
27		0	6	
28		2	20	

In C27, enter the formula =B27*B27+5*B27+6. The program calculates the formula and returns 6. Copy the formula to C28. The program returns the value 20.

Access SOLVER from the toolbar. Use TOOL then SOLVER. A dialog box will appear. The cell address B27 will appear in the dialog box. Under VALUE, click 0. In changing cell, click C27.

Set target cell	$C28
Value	0
By changing cells	$2B27:$B28

The results will be comparable to those shown below.

	A	B	C	D
26		Roots	Equation	
27		3.00E+00	6.00E+00	0
28		2.00E+00	6.00E+00	0

The roots are 3 and 2.

Area of Circle

Formula: A = Pi*R*R

Problem: A clockmaker wants to make a series of interlocking clocks with diameters ranging from 1 to 320 inches. He wants to know how much plywood to purchase.

Diameter	=	A
Radius	=	B
Pi	=	C
Formula for area		=B16*B16*C16

Enter data for the diameter of the clock. Enter 1 and copy to 20.

Calculate radius thus: $\frac{\underline{A16}}{2}$ copy to B29

In C16 enter 3.143. In D16 enter the formula =B16*B16*C16. Copy to D29. Access INSERT from the toolbar. Select FUNCTION and sum D16:D29.

	A	B	C	D
No.	Diameter	Radius	Pi	Area
16	1	0.5	3.143	0.7858
17	2	1.0	3.143	3.1430
18	3	1.5	3.143	7.0718
19	4	2.0	3.143	12.4720
20	5	2.5	3.143	19.6438
21	6	3.0	3.143	28.2870
22	7	3.5	3.143	38.5018
23	8	4.0	3.143	50.2880
24	9	4.5	3.143	63.6458
25	10	5.0	3.143	78.5750

26	11	5.5	3.143	95.0758
27	12	6.0	3.143	113.1480
28	13	6.5	3.143	132.7918
29	14	7.0	3.143	154.0070
Total				797.5366

C. Trigonometry

Area of Triangles

Objective: Computing the areas of triangles using trigonometric functions.

Problem: The engineer wants to calculate the area of a number of triangles for a bridge. He has decided on the dimensions of two sides and the corresponding interior angles of each triangle. John decides to use the sine function.

Trigonometric functions are expressed in radians. Excel will return the values in radians.

First side A14
Second side B14
Angle: C14
Formula for area: 0.5*A14*B14*Sin(Radians(C14))

	A	B	C	D
No.	Side A	Side B	Angle	Area of Triangle
14	1	5	45	1.76777
15	2	6	47	4.38812
16	3	7	51	8.16003
17	4	8	54	12.94427
18	5	9	59	19.28630
19	6	10	62	26.48840
20	7	11	65	34.89280
21	8	12	68	44.50480
22	9	13	71	55.31280
23	10	14	72	66.57400
24	11	15	80	81.24660
25	12	16	85	95.63470
26	13	17	86	110.23080
27	14	18	88	125.92324

28	15	19	89	142.47829
29	16	20	90	160.00000
30	17	21	90	178.50000
31	1	2	90	1.00000
32	5	5	90	12.50000

Instructions: Enter data as above. Under Side A, enter the dimensions of Side A of the triangles. In A14, enter 1 and copy to A32. Under Side B, enter the dimensions of Side B of the triangles. In B14, enter 5 and continue. In C14, enter the value of the angle and continue. In D1, enter the formula =(0.5*A14*B14*Sin((Radians(C14))). Copy the formula to the other dimensions of the triangles. The table above shows the output.

D. Financial Functions

To calculate payment on a mortgage loan, PMT (payment per month) and IMPT (interest payment per month).

Problem: John plans to get a mortgage on a condominium. The sale price is $100,000. He has to decide whether he should take out a mortgage for 15 years at 7% or 10 years at 8%.

Divide annual rate by 12 to 12 payments per month. The options are:

12 * 15 = 180 payments
12 * 10 = 120 payments

Option 1: Loan of 100,000 for 15 years at 7%.

Step 1: Access the PASTE function on the toolbar. A dialog box will appear. Select FINANCIAL from the menu. Choose PMT and click OK. In the FUNCTION wizard, enter the cell address of the data, thus:

Under Rate, enter C2. The rate will appear:	0.0583
Under Nper, enter C3. The periods will appear:	180
Under Pv, enter C4. The principal will appear:	100,000

Click on OK and the monthly payment (principal and interest) will appear.

	A	B	C	D	E
1	Rate		7%	Function	Wizard
2	Rate per period		0.00583	PMT	Value $898.00
3	# of payments		180	rate	Fx[C2] 0.00583
4	Principal Amount		-100,000	nper	Fx[C3] 180
5	Monthly payment		(C2,C3,C4)	pv	Fx[C4]-100,000
6	Monthly payment		$898.60		

Step 2: To calculate interest in period 1, click on the PASTE function from INSERT. A dialog box will appear. Select IMPT. A dialog box will appear.

Under Rate, enter C2. The rate will appear:	0.00583
Under Per, enter 1. The period will appear:	1
Under Nper, enter C4. The under of periods will appear:	180
Uder Pv, enter C4. The principal will appear:	-100,000

	A	B	C	D	E
1	Rate		7%	Function	Wizard
2	Rate per period		0.00583	IPMT	Value $583.00
3	# of payments		180%	rate	fx[C2] 0.00583
4	Principal		-100,000	1	fx[1] 1
5				nper	fx[C3] 180
6	Interest in period		(C2, 1, C3,C4)	pv	fx[C4] -100,000
6	Interest in period		$583.00		
7	Monthly payment		$305.60		

Option 2: To calculate the payment including principal and interest on loan of $100,000 for 10 years at 8%, access the PASTE function by selecting INSERT, FUNCTION. Select FINANCIAL and them PMT. Complete the form, applying the required data.

1	Rate	8%	Function	Wizard
2			PMT	Value $1,212.85
3	Rate per period	0.00555	rate	fx[C2]0.00583
4	# of payments	120	nper	fx[C3]120
5	Principal	-100,000	pv	fx[C4]-100,000
6	Monthly payment	(C2,C3,C4)		
7	Monthly payment	$1,212.85		

Step 2: To calculate interest payment in period 1, access the PASTE function, then FINANCIAL, then IMPT. Enter data: D6 for rate, 1 for period 1, D65 for # of payments, and D6 for principal.

A	B	C	D	E
1 Rate		8%	Function	Wizard
2 Rate per period		0.00555	IPMT	Value 666.0
3 # of payments		120	rate	fx[C2]0.666
4 Principal		-100,000	per	fx[1] 1
5			nper	fx[C3].120
6 Interest in period		(C2,1,C3,C4)	pv	fx[C4]-100,000
7 Interest in period		$666.00		

Summary of Payments on Loan of		$100,000
	15 Years	10 Years
Monthly payment	$898.60	$1,212.84
Time	15 years	10 years
Rate	7%	8%
Total payment	$161,748	$145,542

Which option did John choose?

Partial Amortization Schedule		
Rate	7%	8%
Rate per period	0.00583	0.00666
# of payments	180	120
Principal	-100,000	-100,000

Mon. Pay. Interest	898.6 Principal	Mon. Pay. Interest	1212.85 Principal
583.00	315.40	666.00	546.85
581.16	317.44	662.36	550.49
579.32	319.28	658.72	557.77
577.48	321.12	655.08	557.77
575.64	322.96	651.44	561.41
573.80	324.80	647.80	565.05
571.96	326.40	644.16	568.69
570.12	328.48	640.52	572.33
568.28	330.32	636.88	575.97
566.44	332.16	633.24	579.61
564.60	332.00	629.60	583.25
562.76	335.84	625.96	586.89
560.92	337.60	622.32	690.18
559.08	339.52	618.68	594.17
557.24	341.36	615.04	579.81

555.40	343.24	611.40	601.45
553.56	345.04	607.76	605.09
551.72	346.88	604.12	608.73
549.88	348.72	600.48	612.37
548.04	350.56	596.84	616.01
546.20	352.40	593.20	619.65
544.36	354.24	589.56	623.29
542.52	356.08	585.92	626.93
540.68	357.92	582.28	630.57
538.84	359.76	578.64	634.21
537.00	361.60	575.00	637.85
535.16	363.44	571.36	641.49
531.48	367.12	564.08	648.77
529.64	368.96	560.44	652.41

Simple Interest Versus Compound Interest

Problem: James and John had $5000 each. John deposited his money in an account that paid 10% at simple interest and John invested his money in a certificate of deposit that paid 10% at compound interest. What amounts did James and John have after 5 years?

Formula for simple interest: S1 = P * T * R
Formula for compound interest: Y = B * N
In H24, enter the formula: =(E24*(EXP(F24*G24)))

Copy the formula form H24:H33. The output is shown below.

	A	B	C	D	E	F	G
	James Initial Dep.	Time	Rate @10%	John Initial Prin.	Rate @10%	Time	Amount
24	5000	1	0.10	5000	0.10	1	5525.855
25		2		5000	0.10	2	6107.014
26		3		5000	0.10	3	6749.294
27		4		5000	0.10	4	7459.123
28		5		5000	0.10	5	8243.606
29		6		5000	0.10	6	9110.594
30		7		5000	0.10	7	10068.764
31		8		5000	0.10	8	11127.705
32		9		5000	0.10	9	12298.016
33		10		5000	0.10	10	13591.016
34							

Simple interest on $5,000 at 10% for 10 years:	4000.00
Amount:	9000.00
Amount on $5000 for 10 years at 10%:	12,591.01

Note: Banks generally compound interest on accounts.

5525.855
6107.014
6749.294
7459.123
8243.606
9110.594
10068.764
11127.705
12298.016
13591.016

E. Statistics

Problem: The Teacher Education Department selects students on the basis of their perceived abilities. To be admitted, a student must have an initial grade point average (GPA) of 2.70. The Department wanted to know whether GPA and age could predict performance. Regression analysis was used to examine the problem. The dependent variable was final GPA and the independent variables were initial GPA and age.

Task 1: Input data.

A Student	B Initial GPA	C Age	D Final GPA
1	2.70	21	3.0
2	2.75	26	4.0
3	3.90	20	3.9
4	2.80	25	4.0
5	3.00	20	3.9
6	3.60	21	4.0
7	2.80	25	3.9
8	3.70	26	3.8
9	3.90	30	3.9
10	2.70	30	2.7
11	3.50	35	3.0
12	2.78	36	3.0

13	2.66	38	3.9
14	2.87	30	3.0
15	3.90	25	3.9
16	3.60	38	3.9
17	3.00	27	3.0
18	2.88	25	3.4

Task 2: Highlight data. Access TOOLS, then DATA ANALYSIS. Select REGRESSION.

In Input Range Y, enter D1:D18
In Input Range X, enter A1:B18
In Output Range, enter F19:F40 (or select other range).

Complete all options and click OK. The output is shown on the following page.

Interpretation of Regression

Multiple R of .65 indicates that 65% of the variance in performance is explained by the independent variables, initial GPA and age. The proportion of variation is a number in a range from zero to one and is

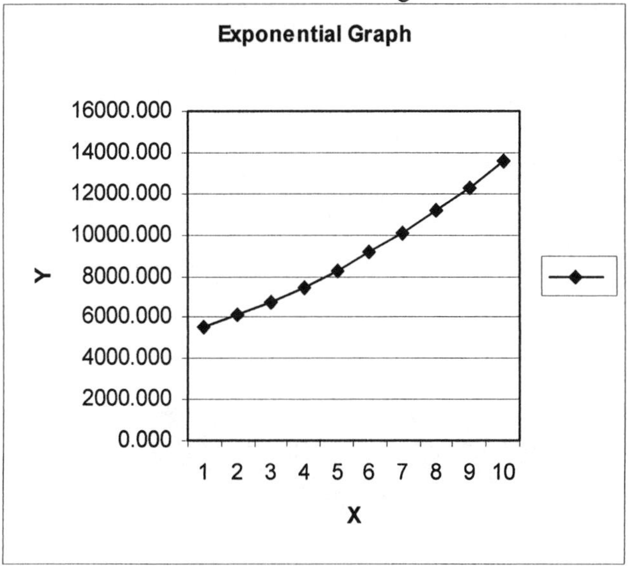

expressed as a percentage. However, R-Square more accurately indicates the amount of variation in the dependent variable than is explained by the input variables.

	Residual Output		Probability Output	
	Predict GPA	Residuals	Percentile	Final GPA
1	2.960	0.040	2.778	2.70
2	3.100	0.900	8.333	3.00
3	4.082	-0.182	13.889	3.00
4	3.129	0.871	19.444	3.00
5	3.226	0.674	25.000	3.00
6	3.815	0.185	30.556	3.00
7	3.129	0.771	36.111	3.40
8	4.004	-0.204	41.667	3.80
9	4.269	-0.369	47.222	3.90
10	3.128	-0.428	52.778	3.90
11	3.982	-0.982	58.333	3.90
12	3.316	-0.316	63.889	3.90
13	3.239	0.661	69.444	3.90
14	3.289	-0.289	75.000	3.90
15	4.175	-0.275	80.556	3.90
16	4.133	-0.233	86.111	4.00
17	3.357	-0.357	91.667	4.00
18	3.205	0.195	97.222	4.00
	2.795	0.605	97.222	4.00

ANOVA

	df	F	Significance
Regression	2	1.8582	
Residual	16		
Total	18		

	Coefficients	P-value	Lower 95%	Upper 95%
Intercept	0			
Init. GPA	0.950874707	1.675E-05	0.61787778	1.2838716
Age	0.018676387	0.3097874	-0.0190681	0.0564209

Coefficients

The null hypothesis that there is no relationship between the dependent variable, performance, and the independent variables, initial GPA and age. The coefficient 0.95 indicates that 95% of the variance in performance may be predicted by GPA, and the coefficient 0.01 indicates that age predicts only 1% of the final GPA.

Interpretation of Analysis of Variance

The probability of getting these results in a random sample is indicated by the statement of significance. It is 1.8.

Predicted Results

The grades of individual students can be predicted by comparing the residual output and the probability output.

Appendices

Appendix A Chronological Development of Mathematics to Technology

Appendix B Development of Numerals

- Babylonian
- Egyptian
- Hebrew Numerals
- Mayan
- Chinese
- Hindu
- Arabic
- Modern European

Appendix A

Chronological Development of Mathematics to Technology

B.C.E.

4700	Early beginning of Babylonian Calendar
4241	Egyptian calendar introduced
3000	Babylonian civilization, Hammurabi 2100
3000	Egyptian civilization
2900	Building of the Great Pyramid
2852	First Emperor of China, Fuh-hi Beginning of Chinese astronomy
1650	Ahmes (The Rhind Papyrus written)
1500	Egyptian sundial Mathematics applied to taxation.
1350	Rameses in Egypt
1122	Chow Dynasty in China; growth of mathematics
1055	King David of Israel ascended the throne
1000	Indian civilization began: the Vedic verses
753	Founding of Rome
660	Japanese numeration, power of 10
550	Ameristus's introduction of geometry
542	China: use of bamboo rods for counting
540	Pythagoras: geometry, number theory
380	Plato: *Foundations of Mathematics*
340	Aristotle: logic and application of mathematics
336	Alexander the Great: Beginning of the Greek Empire
300	Euclid: geometry
247	Ptolemy: astronomy, geometry, trigonometry
225	Apollonius: conics Archimedes: geometry

140	Hipparchus: trigonometry, astronomy
44	Caesar assassinated on the Ides of March
C.E.	
1	Rag paper in China
26	Rise of Christianity in Judea
44	Augustus and the rise of the Roman Empire
66	New Calendar in China
120	Hyginus and surveying
150	Claudius Ptolemy: astronomy, geometry, trigonometry
275	Diophantus: algebra
300	Pappus: geometry
313	Christianization of the Roman Empire by Constantine
410	Hypatia of Egypt: astronomy, geometry
476	Fall of the Roman Empire
510	Boethus, Anicius: geometry
	Aryabhata of of Kusumapura: algebra
622	Mohammed flees from Mecca, beginning of Islam
625	In China, Wang Hsiao-tung: cubic equations
628	Brahmagupta of Gwalior:Geometry and algebra
642	Burning of the Library at Alexandria
650	Sebokht: Hindu numerals
820	Al Khowarizmi, Mohammed ibn Musa: algebra
870	Tabit ibn Qurra: Greek mathematics
1076	Turks captured Jerusalem
1095	The First Crusade
1100	Omar Khayyam: algebra, geometry
1175	Averroes: astronomy, trigonometryt
1202	Leonard of Pisa, Fibonacci: algebra, geometry
1225	Jordanuus Nemorarius: algebra
1250	Rise of Universities in Europe
	Roger Bacon: astronomy
1453	Fall of Constantinople and beginning of the Renaissance
1470	Regiomontanus: trigonometry
1500	Leonardo da Vinci: geometry, optics
1506	Scipione del Ferro: algebra, cubic equation
1542	Robert Recorde: algebra, geometry
1543	Copernicus, Nicholas: Heliocentric theory of the Universe
1564	Galileo Galilei: astronomy, Heliocentric theory of the Universe
1642	Isaac Newton: calculus, gravitation

1572 Bombelli: algebra
1580 Francois Viete: algebra
1610 Johann Kepler: astronomy, geometry
1614 John Napier: logarithm
1635 Pierre de Fermat: analytic geometry, number theory
1637 Rene Descartes: analytic geometry
1650 Blaise Pascal: geometry, probability, number theory, mechanical adding machine with rotating wheels
1670 Christian Huygens: geometry, physics, astronomy
1670 Isaac Newton: calculus
1678 Edmund Halley: astronomy
1682 Gottfried Wilhelm Leibniz,: calculus
1694 Gottfried Wilhelm Leibniz: calculating machine that multiplied by repeated addition
1690 Marquis de L'Hospital: applied calculus
1700 Jacques, Jean and Daniel Bernoulli: Applied calculus
1750 Leonard Euler: analysis, physics, astronomy
1799 Pierre-Simon LaPlace: celestial mechanics, astronomy, probability, differential equations, calculus
1780 Joseph Larange, Joseph: number theory, analysis, elliptical functions
1800 Carl Gauss: number theory, analysis, general mathematics
1825 Abel: elliptical functions
1830 Charles Baggage: calculating machine
1832 Leopold Kronecker: elliptical functions, equations, topology
1850 William Hamilton: quaterions
Salmon De Morgan: algebra, logic
George Bolle: logic, differential equations
1850 Ernst Eduard Kummer: series, complex numbers, cubic equations
1850 George Riemann and Friedrich Bernhard: surfaces, elliptical functions
1850 Ferdinand Einstein, Max Gotthold: invariants
Gudermann: hyperbolic functions
1869 Georg Cantor: Theory of Assemblages
1872 James Joseph Sylvester: algebra, invariants
1880 Herman Hollerich: punched-hole machine for calculation
1896 Herman Hollerich: founding of the Tabulating Machine Company
1910 Bertrand Russell, Alfred Whitehead: symbolic logic
1914 Founding of International Business Machines (IBM)

1940	Development of Mark I computer
1940	John Atanasoff: binary base machine calculations
1942	John Atanasoff: Atanasoff Berry Computer (ABC), a working computer
1942	John Mauchly, Presper Eckert: Electronic Numerical Integrator and Calculator (ENIAC)
1949	Cambridge University, England: Electronic Delay-Storage Automatic Computer (EDSAC), a complete stored-program computer
1951	Sperry Rand: UNIVAC at U.S. Bureau of the Census
1954	IBM: first commercial computer in Boston
1975	Bill Gates: Founding of Microsoft

Appendix B

Development of Numerals

This appendix describes the historical development of the numerical system from the earliest beginnings in Babylonia to its flowering in early modern Europe. The references include H. L. Resnikoff and R. O. Wells (1973), and D. E. Smith (1923, 1953).

Babylonian Numerals

The Babylonian writing was wedge-shaped and was written on clay tablets. It was a sexagesimal system with a base of sixty that was suitable for scientific purposes. There were two basic notations, which correspond to one (𒁹)and ten (𒌋), similar to those used in the Egyptian decimal system.

Examples:

1 𒁹
3 𒁹𒁹𒁹
25 𒌋𒌋 𒐊

Because the symbols had a base of sixty, the symbols had different numerical meanings. For example, 𒁹 not only meant one but also meant multiples of sixty. For instance:

$2.60 + 23 = 143$ 𒁹𒁹𒌋𒌋𒁹𒁹𒁹

The two symbols could be used to represent fractions. Each time a new place value would appear.

$1.60^2 + 1.60 + 3$ 𒁹 𒁹 𒁹𒁹𒁹

Egyptian Numerals

Hieroglyphics is the oldest form of the Egyptian written language. They were used to record information on monuments and tombs. The Egyptians used a partially positional decimal system that was based on counting by powers of ten rather than by position. Hieroglyphics symbols were written as follows:

 One

 Ten

 Hundred

 Thousand (lotus flower)

 Ten thousand

 Hundred thousand

 One million ("god of the infinite")

 Ten million

Hieroglyphics Numerals

Normally the numbers were written from left to right. The values were added by representing the values by the symbols.

Later Egyptian writing was in cursive script called *Hieratic*. In modern times the writing is called *Demotic*. Knowledge of Egyptian writing is derived from two documents written on papyri: the first is named the Rhind Papyrus, named for the finder, A. Rhind. It is now in the British Museum. The other document, the Moscow Papyrus, is now in the Museum of Fine Arts, Moscow. These documents were written in cursive script between 2060 and 1580 B.C.E.

Hieratic Numerals

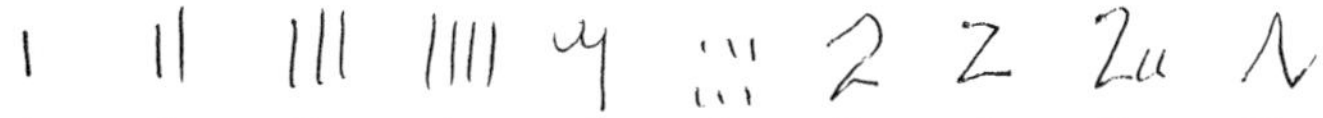

1 2 3 4 5 6 7 8 9 10

Typical Notations

	½	1	10	100	1000
Hieroglyphics					
Hieratic					
Demotic					

Hebrew Numerals

א	ב	ג	ד	ה	ו	ז	ח	ט
1	2	3	4	5	6	7	8	9
י	כ	ל	מ	נ	ס	ע	פ	צ
10	20	30	40	50	60	70	80	90
ק	ר	ש	ת	ך	ם	ן	ף	ץ
100	200	300	400	500	600	700	800	900

Mayan Numerals

The Babylonian number system ranged on a scale from to 20 and had two components. A dot (*) represented 1 and a dash (-) represented 5. The system also had a zero represented by (). The first twelve numerals appeared as follows:

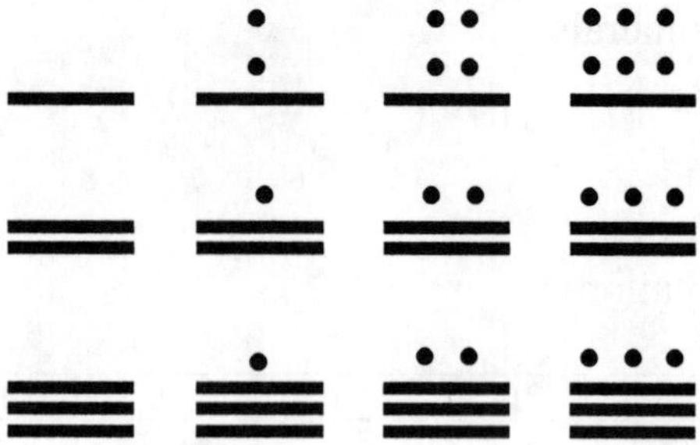

Chinese Numerals

Earliest form of Chinese numerals used by merchants was as follows:

During the Sung Dynasty (950–1280), numbers were represented by rods, with the zero appearing as a circle.

Contemporary form of Chinese numerals from 1 to 10 may be written in descending order, that is, from the top downward, or they may be written from left to right.

一 二 三 四 五

六 七 八 九 十

Hindu Numerals

Early forms of Hindu numerals are shown below.

1 2 4 6 50 50 200 200

Later modern Sanskrit numerals, and the later form, which included the zero are shown below:

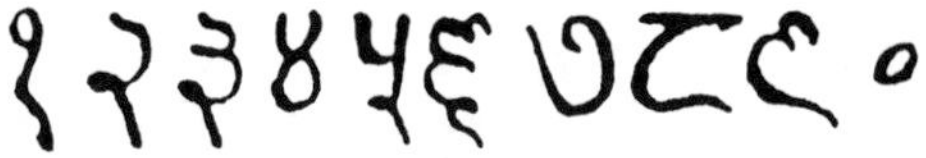

Greek Numerals

1	α	alpha	60	ξ	xi
2	β	beta	70	ο	omicron
3	γ	gamma	80	π	pi
4	δ	delta	90	ς	koppa
5	ε	epsilon	100	ρ	rho
6	ς	vau	200	σ	sigma
7	ζ	zeta	300	τ	tau
8	η	eta	400	υ	upsilon
9	θ	theta	500	φ	phi
10	ι	iota	600	χ	chi
20	κ	kappa	700	ψ	psi
30	λ	lambda	800	ω	omega
40	μ	mu	900		sampi
50	ν	nu	1000	.α	

Roman Numerals

I	1
II	2
III	3
IV	4
V	5
V1	6
V11	7
V111	8
1X	9
X	10

X1	11
L	50
C	100
M	1000
$\overline{\text{M}}$	1,000,000

Arabic Numerals

Arabic numerals are similar to the Sanskrit form of the Indian numerals with the inclusion of the zero.

Sanskrit

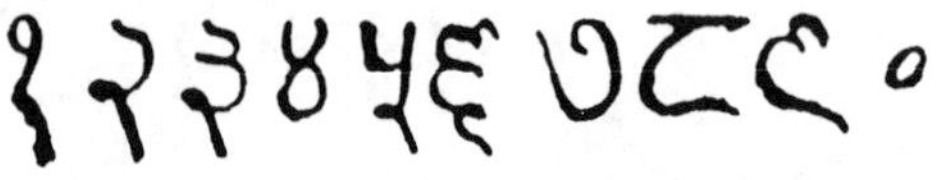

Arabic

Early Modern European Numerals

This form of the numerals was first seen in Europe in 976, but variations have been seen in Arab manuscripts a century later dated 874 and in an Egyptian manuscript in 961. It is generally believed that the numerals were derived from Moorish mathematics. The numerals were used by Gerber and it is believed that he obtained it from a Spanish manuscript dated 976. This manuscript may have been based on knowledge obtained from mercantile activities of traders with the East rather than from Arab sources. The earliest use of the numerals was seen on a Sicilian coin in 1138 (Smith, 1953).

Endnotes

Chapter 1

Baumgart, J. (1989.) History of Algebra. In J. Baumgart, *Historical Topics for the Mathematics Classroom* (p. 243). Washington, DC: National Council of Teachers of Mathematics.

Cooke, R. (1993). *The History of Mathematics, A Brief Course* (pp. 183, 241, 247, 268). New York: John Wiley and Sons, Inc.

Eves, H. (1989). The History of Geometry. In *Historical Topics for the Mathematics Classroom* (pp. 165-230). Washington, DC: National Council of Teachers of Mathematics.

Nelson, D. J., and William, J. (Eds.). (1995). *Multicultural Mathematics: Teaching from a Global Perspective.* www.ed.gov/databases/ERIC_Digests

Struik, D. (1987). *A Concise History of Mathematics* (pp. 212). New York: Dover Publications.

Wrestler, F. E. (1989). Capsule 79. In J. K. Baumgart, *History of Algebra* (pp. 232-332). Washington, DC: National Council of Teachers of Mathematics.

Chapter 2

Brademas, J. (1987). *The Politics of Education* (p. 9). Norman: University of Oklahoma Press.

Bereday, G. Z. F. (1964). *Comparative Methods in Education.* New York: Holt, Rinehart, and Winston, Inc.

Cangelosi, J. (1906), *Teaching Mathematics in Secondary School: An Integrative Approach* (p. 404). Englewood Cliffs, NJ: Prentice Hall.

Carlson, R. (1996). *Reframing and Reform* (pp. 199-201). White Plains, NY: Longman.

Cooke, R. (1997). *The History of Mathematics: A Brief Course* (pp. 466-457). New York: John Wiley & Sons, Inc.

Gutek, G. (1995). *A History of Western Experience* (pp. 434, 514, 518). Prospect Heights, IL: Waveland Press.

Huetinch, L., and Munshin, S. (2000), *Teaching Mathematics for the 21st Century* (pp. 5, 13, 218). Upper Saddle River, NJ: Merrill.

National Council of Teachers of Mathematics. (2000). *Curriculum and Evaluation Standards, 2000.* Washington, DC: Author.

Polya, G. (1985). *How to Solve It: A New Aspect of Mathematical Method* (p. 5). Princeton, NJ: University Press.

Struik, D. (1987). *A Concise History of Mathematics* (p. 210). New York: Dover Publications.

United States Department of Education. (2000). *The Condition of Education* (pp. 76, 224). Washington, DC: Government Printing Office.

Urban, W., & Wagoner, J. (2000). *American Education: A History* (p. 298). New York: McGraw Hill.

Chapter 3

Brademas, J. (1987). *The Politics of Education* (p. 9). Norman: University of Oklahoma Press.

Carlson, R. (1996). *Reframing and Reform* (p. 199-201). White Plains, NY: Longman.

Cremin, L. (1961). *The Transformation of the School: Progressivism in American Education 1876-1957*(p. 24). New York: Knopf.

DeWitt, N. (1962). *Professional Manpower* (pp. 255-257). Washington, DC: National Science Foundation.

Gutek, G. (1995). *A History of Western Experience* (pp. 434, 514, 518) Prospect Heights, IL: Waveland Press.

Hans, N. (1964). *Comparative Education: A Study of Educational Factors and Tradition.* London: Routledge and Kegan Paul.

Havighurst, J. (1958, April 26). Is Russia Really Out-performing U.S. in Scientists? In *School and Society*, 187-192.

Johnson, W. H. (1950). *Russia's Educational Heritage.* Pittsburgh, PA: Carnegie Institute of Technology.

Mallinson, V. (1964). *Comparative Education* (p. 248). London: Heineman.

Nikandrov, N. (2000). Education in Modern Russia: Is it Modern? In K. Mazurek, M. Winzer, and C. Majorek (Eds.), *Education in a Global Society* (pp. 209-223). Boston: Allyn and Bacon.

Struik, D. (1987). *A Concise History of Mathematics* (p.210). New York: Dover Publications.

Timoshenko, S. P. (1959). *Engineering Education in Russia* (p. 5). New York: McGraw Hill.

Chapter 4

Desmonds, E. (1995). The Failed Miracle. *Time Magazine* (pp.60-64).

Goya, S. Japanese Education: Hardly Known Facts. In *The Education Digest, 58*, 8-12.

Heutinch, L., & Munshin, S. (2000). *Teaching mathematics for the 21st Century* (pp. 5, 13, 218). Upper Saddle River, NJ: Merrill.

Kanaya, T. *International Encyclopedia of National Systems of Education* (2nd ed., pp. 482-488), T.N. Postlewaite (Ed.). New York: Elsevier Science).

McNergney, R., & Herbert, J. (1998). *Foundations of Education: The Challenge of Professional Practice*. Boston: Allyn and Bacon.

Stevenson, H., & Stigler, J. (1992). *The Learning Gap: Why Our Schools Are Failing and What We Can Learn from the Japanese and Chinese Education in American Education*. New York: Summit.

United States Department of Education. (2000). *The Condition of Education*. Washington, DC: Government Printing Office.

Chapter 5

Gredler, M. (1986). *Learning and Instruction: Theory into Practice* (pp. 14-21). NewYork: McMillen Publishing.

Frost, J. (1962). *Basic Teaching of the Great Philosophers* (pp. 117, 222). New York: Anchor Books.

Hamlyn, D. (1987). *History of Western Philosophy* (pp. 134-168). New York: Penguin Books.

Henderson, J. (1996). *Reflective Teaching* (p. 16). Columbus, OH: Merrill-Prentice Hall.

Koestenbaum, P. (1968). *Philosophy: A General Introduction* (pp. 157-253). New York: Van Nostrand Reinhold Co.

Chapter 6

Atkinson, R. C., & Schiffer, R. M. (1968). Human Memory: A Proposed System and Its Control Process. In K. Spence and J. Spence (Eds.), *The Psychology of Learning and Motivation.* New York: Academic Press.

Beihler, R., & Snowman, J. (1997). *Psychology Applied to Teaching* Boston: Houghton Mifflin Company.

Bell-Gredler, M. (1986). *Learning and Instruction: Theory into Practice* (pp. 16, 22). London: MacMillan.

Boring, E. G. (1950). *A History of Experimental Psychology.* New York: Appleton-Century-Crofts.

Frost, S. (1962). *Basic Teachings of the Great Philosophers* (p. 30). London: Anchor Books, Doubleday.

Gagne, Robert. (1985). *The Conditions of Learning and Theory of Instruction.* Chicago: Holt, Rinehart, and Winston.

Hamlyn, D. (1987). *Western Philosophy* New York: Penguin Books.

Koffka, K. (1935). *Principles of Gestalt Psychology* (p. 141). New York: Harcourt Brace.

Lewin, K. (1936). *Principles of Tolological Psychology.* New York: Harcourt Brace.

Lovell, K. (1973). *Educational Psychology and Children* (pp. 122). London: Hodder and Stroughton.

Smith, G., & Smith, J. (1994). *Education Today* (pp. 155). New York: St. Martin's Press.

Tolman, E. (1948). *Collected Papers in Psychology.* New York: Harcourt Brace.

Chapter 7

Hannaford, C. (1995). *Smart Moves.* Newport News, VA: Great Ocean Publishing.

Hobson, J. (1994). *Chemistry of Conscious States.*(Boston: Little Brown and Co.

Howard, P. (1994). *Owner's Manual for the Brain.* Austin, TX: Leornian Press.

Hutchinson M. (1994). *MegaBrain Power.* New York: Hyperion Books.

Jensen, E. (1998). *Teaching with the Brain in Mind* Alexandria, VA: Association for Supervision and Curriculum Development.

Kesslyn, S. (1992). *Wet Mind.* New York: Simon and Shuster.

LeDoux, J. (1996). *The Emotional Brain.* New York: Simon and Shuster.
Ostrander, S., & Schroeder, L. (1991). *SuperMemory.* New York: Carrol and Graf Publishers.
Vincent, J. F. (1990). *The Biology of Emotions*. Cambridge, MA: Basil Blackwood.

Chapter 8

Blair, T. (1988). *Emerging Patterns of Teaching: From Methods to Field Experiences* (pp. 51-56) Columbus, OH: Merrill Publishing.
Gagne, R. (1985). *The Conditions of Learning* (pp. 70-105). New York: Holt, Rinehart, and Winston.
Glaser, R. (1990). The Emergence of Learning Theory Within Instructional Research. *American Psychologist, 45*, 29.
McClelland, D. (1968). The Urge to Achieve. *Psychology of Organization.*
Rosenshine, B., and Stevens, R. (1986). Teaching Functions. In M. Wittrock (Ed.), *Handbook of Research on Teaching* (pp. 377-390). New York, MacMillan.
Smith, G and Smith J. *Education Today* (New York, St. Martin's Press: 1994) p.149-159.

Chapter 9

Colangelo, J. (1996). *Teaching Mathematics in Secondary Schools* (pp. 24-176). Englewood Cliffs, NJ: Prentice Hall.
Gagne, R. (1985). *The Conditions of Learning and Theory of Instruction* (p. 342). Fort Worth, FL: Holt, Rinehart and Winston Inc.
Maslow, A. (1964). *Hierarchy of Needs.* New York: Harper and Row.
National Council of Teachers of Mathematics. (2000). *Curriculum and Evaluation Standards, 2000.* Washington, DC: Author.
Stigler, J., & Stevenson, H. (1991, Spring). How Asian Teachers Polish to Perfection. *American Educator*, pp. 31-38.
Tyler, R. (1993). *Basic Principles of Curriculum and Instruction.* Chicago: The Open University Press.
White, R. W. (1959). Motivation Reconsidered: The Concept of Competence. *Psychological Review, 66*, 297-333.
Willoughby, S. (1990). *Mathematics for a Changing World* (pp. 1-8). Alexandria, VA: Association for Supervision and Curriculum.

Chapter 10

Bloom, B. S. (Ed.). (1999). *Taxonomy of Educational Objectives, Book 1: Cognitive Domain.* New York: David McKay.

Colangelo, J. (1996). *Teaching Mathematics in Secondary Schools.* Englewood Cliffs, NJ: Prentice Hall.

Gagne. R. (9185). *The Conditions of Learning: Theory of Instruction.* Fort Worth, FL: Holt, Rinehart and Winston, Inc..

Maslow, A. (1964). *Motivation and Personality.* New York: Harper and Row.

Posamentier, A., and Carpenter, J. (1999). *Teaching Secondary School Mathematics.* Upper Saddle River, NJ: Merrill, Prentice Hall.

Tyler, R. (1993). *Basic Principles of Curriculum and Instruction.* Chicago: The Open University Press,.

Willoughby, S. (1990). *Mathematics for a Changing World* (pp. 1-8). Alexandria, VA: Association for Supervision and Curriculum.

Chapter 11

Blair, F. (1998). *Emerging Patterns of Teaching: From Methods to Field Experience* (pp. 51-56, 155-172). Columbus, OH: Merrill Publishing.

Bloom, B. S. (Ed.). (1999). *Taxonomy of Educational Objectives, Book 1: Cognitive Domain.* New York: David McKay.

Brubacher, J. (1969). *Modern Philosophies of Education* (Ch. 8, pp. 238-246). New York: McGraw Hill.

National Council of Teachers of Mathematics. (2000). *Curriculum and Evaluation Standards, 2000.* Washington, DC: Author.

Chapter 12

Beihler, R., & Snowman, J. (1997). *Psychology Applied to Teaching* Boston: Houghton Mifflin Company.

Brubacher, J. (1969). *Modern Philosophies of Education* (Ch. 8, pp. 238-246). New York: McGraw Hill.

Bruner, J. (1960). *The Process of Education* (pp. 6-9). Boston: Harvard University Press.

Gagne, Robert. (1985). *The Conditions of Learning and Theory of Instruction.* Chicago: Holt, Rinehart, and Winston.

Kennedy, L. M., & Tipps, S. (1994). *Guiding Children Learning Mathematics.* Belmont, CA: Wadsworth Publishing Co.

Stevenson, H., & Stigler, J. (1992). *The Learning Gap: Why Our Schools are Failing and What We Can Learn from the Japanese and Chinese Education in American Education.* New York: Summit.

Chapter 13

Anderson, G. J. (1970). Effects of Classroom Social Climate on Individual Learning. *American Education Research Journal, 7*, 135-152.

Axelrod, S. (1977). *Behavior Modification for the Classroom Teacher.* (New York: McGraw Hill.

Brubacher, J. (1969). *Modern Philosophies of Education* (Ch. 8, pp. 238-246). New York: McGraw Hill.

Canter, L., & Canter, M. (1987). *Assertive Discipline* (p. 62). Santa Monica, CA: Canter and Canter Associates, Inc.

Cohen, E. (1999). Making Cooperative Learning Equitable. *Educational Leadership, 56*, 18-21.

Dreikurs, R., & Cassel, P. (1972). *Discipline Without Tears*. New York: Hawthorn.

Dreikurs, R. (1986). *Psychology in the Classroom* New York: Harper and Row.

Flanders, N. Diagnosing Classroom Social Structure. In W. Elena and B. Biddle (Eds.), *Contemporary Research on Teacher Effectiveness*. New York: Holt, Rinehart and Winston.

Ladoucer, R., & Armstrong. (1983). Evaluation of a Behavior Program for the Improvement of Grades Among High School Students. *Journal of Counseling, 30*, 100-103.

Kagan, D. M. (1990). How Schools Alienate Students at Risk: A Model for Examining Classroom Variables. *Educational Psychologist, 25*, 102-125.

Johnson, D., Johnson, R., & Holubec, E. (1994). *The New Circles of Learning: Cooperation in the Classroom and School*. Alexandria, VA: Association for Supervision and Curriculum Development.

Kounin, J. (1970). *Discipline and Group Management in the Classroom.* New York: Holt, Rinehart, and Winston.

Kresh, D. Crutchfield, R, & Belarchy, E. (1962). *Individual, and Society.* New York: McGraw Hill Inc.

Orlick, D., Harder, R., Callahan, R., & Gibson, H. (1988). *A Guide to Better Instruction.* Boston: Houghton Mifflin.

Sharon, Y., & Sharon, S. (Eds.). (1994). Group Investigation in the Cooperative Classroom. In *Handbook of Cooperative Learning Methods*. Westport, CT: Greenwood Press.

Skinner, B.F. (1948). *Walden Two.* New York: MacMillan.

Slavin, R. (1995). Cooperative Learning: Theory, Research, and Practice. Boston: Allyn and Bacon.

Walberg, H., & Anderson, A. (1974). *Evaluating Educational Performance.* Berkeley, CA: McCrutchen Publishing Corporation.

Chapter 14

Boyer, E. (1983). *High School: A Report on Secondary Education in America* (pp. 71-202). New York: Harper Colopon Books.

Brubacher, J. (1964). *Modern Philosophies of Education* (pp. 155-172). (Tokyo: Kogakusha.

Dewey, J. (1938). *Democracy and Education.* New York: Colin McMillan.

Taba, H. (1962). *Curriculum Development: Theory and Practice* (pp. 290). New York: Harcourt, Brace, and Wood.

Travers, P. and Rebore, R. (1995). *Foundations of Education.* Boston: Allyn and Bacon.

Tyler, R. *Basic Principles of Curriculum for Instruction* (Chicago: The University of Chicago Press, 1993) pp.3-37.

Webb, L., Mertha, A., & Jordan, K. (2000). *Foundations of Education.* Columbus, OH: Merrill.

Chapter 15

Lewis, D. (1994). *Assessment in Education.* London: University Press of London.

Chapter 16

Bobbitt, F. (1918). *The Curriculum.* Boston: Allyn and Bacon.

Bolin, F. (1995). Teaching as a Craft. In A. Ornstein, *Teaching: Theory into Practice.* Boston: Allyn and Bacon.

Bolin, F., & Panaritis, O. Searching for a Common Purpose: A Perspective on the History of Supervision. In C. Glickman (Ed.), *Supervision in Transition* (pp. 30-43). Alexandria, VA: ASCD.

Dewey, J. (1904). The Relation of Theory to Practice in Teaching. In C. Murray (Ed.), *The Relation of Theory to Practice in Education of Teachers* (pp. 9-30), *Third Yearbook of the National Society for the Scientific Study of Education, Part I*. Chicago: University of Chicago Press.

Eby, J. *Reflective Planning, Teaching and Evaluation K-12* (p. 283). Columbus, OH: Merrill-Prentice Hall.

Henderson, J. (1996). *Reflective Teaching* (p. 16). Columbus, OH: Merrill-Prentice Hall.

Herman, J., Aschbacher, P., & Winters, P. (1992). *A Practical Guide to Alternative Assessment*. Alexandria, VA: Association for Supervision and Curriculum Development.

Rice, J. (1893). *The Public School System of the United States*. New York: Century.

Tom, A. (1984). *Teaching as a Moral Craft*. New York: Longman.

Chapter 17

Ary, D., Jacobs, L., & Razavieh, A. (1992). *Introduction to Research in Education*. New York: Holt, Rinehart and Winston.

Gall, M., Borg W., & Gall J. (1996). *Educational Research*. New York: Longman.

Kerlinger, F. (1976). *Educational Research*. New York: Holt, Rinehart, and Winston.

Persaud, G., & Figueroa, P. (1972). *Sociology of Education*. Oxford, MA: Oxford University Press.

Tuckman, B. (1972). *Measuring Educational Outcomes*. New York: Harcourt Brace and Wood.

Chapter 18

Bohl, Marilyn. (1984). *Information Processing*. Chicago: Science Research Associates Inc.

Heutinch, L., & Munshin, S. (2000). *Teaching mathematics for the 21st Century* (pp. 5, 13, 218). Upper Saddle River, NJ: Merrill.

Russell, A., & Whitehead, R. (1904). *Principia Mathematica*.

Struik, D. (1987). *A Concise History of Mathematics*. New York, Dover Publications.

United States Department of Education. (2000). *The Condition of Education* (pp. 76, 224). Washington, DC: Government Printing Office.

Willoughby, S. (1990). *Mathematics for a Changing World* (pp. 1-8). Alexandria, VA: Association for Supervision and Curriculum.

Bibliography

Anderson, G. J. "Effects of Classroom Social Climate on Individual learning." In *American Education Research Journal* 7 (1970): 135–152.

Armstrong, D. and T. Savage. *Teaching in the Secondary School.* Upper Saddle River, New Jersey: Merrill, Prentice Hall, 2002.

Bell-Gredler M. *Learning and Instruction: Theory into Practice.* New York: McMillan Publishing, 1986.

Biehler, R. and J. Snowman *Psychology Applied to Teaching.* New York: Houghton Mifflin Company, 1997.

Black, J., K. Isaacs, B. Anderson, B. Alcantrara, and Greenough. "Learning Causes Synaptogenesis, While Motor Activity Cues Angiogenesis in Cerebellar Cortex of Adult Rats." *Proceedings of the National Academy of Sciences* (1987).

Bloom, Benjamin. *Taxonomy of Educational Objectives: Cognitive Domain.* New York: David McKay, 1999.

Borich, G. and M. Tombari. *Educational Psychology*. (Reading, Pennsylvania: Longman, 1999.

Brahier, Daniel. *Teaching Secondary and Middle School Mathematics.* Needham Heights, New Jersey: Allyn and Bacon, 2000.

Brubacher, J. *Modern Philosophies of Education.* New York: McGraw Hill, 1964.

Cangelosi, James. *Teaching Mathematics in Secondary School: An Integrative Approach.* Englewood Cliffs, New Jersey: Prentice Hall, 1996.

Charles, C. and K. Barr. *Building Classroom Discipline.* New York: Longman, 1989.

Cooke, Roger. *The History of Mathematics: A Brief Course.* NewYork: John Wiley & Sons, 1977.

Council of Teachers of Mathematics. *Curriculum and Evaluation Standards, 2000.*

Cremin, L. *The Transformation of the School: Progressivism in American Education, 1876-1957.* New York, Knopf, 1961.

Dauben, Joseph. *Mathematical Perspectives.* New York: Academic Press, 1981.

DeWitt, Nicholas. *Professional Manpower.* Washington, DC: National Science Foundation, 1962.

Diamond, M. "Extensive Cortical Depth Measurements and Neurons Size Increases in Cortex of Environmentally Enriched Rats." *Journal of Comparative Neurology* 131: 357–364.

Dye, Thomas. *Understanding Public Policy.* Englewood Cliffs, New Jersey: Prentice-Hall, 1984.

Eby, J. *Reflective Planning, Teaching, and Evaluation.* Upper Saddle River, New Jersey: Merrill, 1998.

Gagne. R. *The Conditions of Learning: Theory of Instruction.* Fort Worth, Florida: Holt Rinehart and Winston, Inc., 1985.

Garcia, J., E. Spalding, and R. Powell. *Contexts of Teaching: Methods for Middle and High School Instruction.* Upper Saddle River, New Jersey: Merrill Prentice Hall, 2001.

Gardner, H. *Multiple Intelligences: The Theory in Practice*. New York: Basic Books, 1993.

Glaser, W. *Control Theory in the Classroom.* (New York: Harper and Row, 1986.

Gleitman, H. *Studies of Mind and Brain*. New York: Kline, 1982.

Glickman, C. *Supervision in Transition.* New York: Association for Supervision and Curriculum Development ASCD, 1992.

Gohaghan, Kenneth. *Quantitative Analysis for Public Policy*. New York: McGraw-Hill, 2000.

Gutek, G. *Historical and Philosophical Foundations of Education.* Columbus, Ohio: Merrill Prentice Hall, 2001.

Hans, Nicholas. *Comparative Education: A Study of Educational Factors and Tradition*. London: Routledge and Kegan Paul, 1964.

Healy, J. *Endangered Minds: Why our Children Can't Think*. New York: Simon and Shuster, 1990.

Hennessy, J., M. King, T. McClure T., and S. Levine. "Uncertainty as Defined by the Contingency Between Environmental Events and the Adrenocortical Response." *Journal of Comparative Physiology,* (1997): 1447–1460.

Heutinch, Linda, and Sara Munshin. *Teaching Mathematics for the 21st Century*. Upper Saddle River, New Jersey: Merrill Prentice Hall, 1999.

Joyce, B., M. Neil, and M. Shower. *Models of Teaching*. Boston: Allyn and Bacon, 1992.

Kemp, J. *Instructional Design*. Belmont, CA: David Lake Publishers, 1977.

Kline, Morris. *Mathematical Thought from Ancient to Modern Times.* New York: Oxford University Press, 1973.

———. *Mathematics: The Loss of Certainty.* New York: Oxford University Press, 1980.

Kurian, Thomas. *World Data.* New York: A World Almanac Publication, 1993.

Liengme, Bernard. *A Guide to Microsoft for Scientists and Engineers.* New York: John Wiley and Sons, 1997.

Maslow, A. *Motivation and Personality.* New York: Harper and Row, 1964.

McClelland, D. "The Urge to Achieve." In *Psychology of Organization.* New York: McGraw Hill, 1988.

Middletown, Michael. *Data Analysis Using Microsoft.* Albany, New York: Danbury Press, 1987.

Morgan, Bryan. *Men and Discoveries.* London: Cox and Wyman Ltd., 1972.

Mutt, Annmarie. *Statistical Abstract of the World.* Detroit, Michigan: Gale, 2000.

National Center for Educational Statistics. *The Third International Mathematical and Science Study* (TIMSS), 1999.

National Council of Teachers of Mathematics. *Assessment Standards* (NCTM), 2000.

———. *The Growth of Mathematical Ideas, Grades K-12.* Washington DC: National Council of Teachers of Mathematics, 1959.

———. *Historical Topics for the Mathematics Classroom.* Washington, DC: National Council of Teachers of Mathematics, 1993.

Neugerbauer, O. *The Exact Sciences in Antiquity.* Providence, Rhode Island: Brown University, 1970.

Ornstein, A. *Teaching: Theory into Practice.* Boston: Allyn and Bacon, 1995.

Ozman, H. and S. Carver. *Philosophical Foundations of Education.* Upper Saddle River, New Jersey: Merrill, 1999.

Posamentier, A. and J. Carpenter. *Teaching Secondary School Mathematics*. Upper Saddle River, New Jersey: Merrill, Prentice Hall, 1999.

Rersdtak, R. *The Brain*. New York: Bantam Books, 1985.

Resnikoff, H. L. and R. O. Wells. *Mathematics in Civilization.* New York: Holt, Rinehart and Winston, Inc., 1973.

Scott, J. F. *A History of Mathematics: From Antiquity to the Beginning of the Nineteenth Century*. London: Taylor and Francis, 1969.

Sincich, Terry. *Statistics by Example*. San Francisco: Dellen Publishing, 1989.

Slavin, R. "Research on Cooperative Learning: Consensus and Controversy." *Educational Leadership,* 47: 52–54.

Smith, David Eugene. *History of Mathematics*. New York: Dover Publications, 1925.

Smith, Eugene. *History of Mathematics*. Boston: Ginn and Company, 1923.

Smith, G. and J. Smith. *Education Today*. New York: St. Martin's Press, 1994.

Soltis, J. *Philosophy and Education*. Chicago: University of Chicago Press, 1981.

Stigler, J. and H. Stevenson. "How Asian Teachers Polish the Stone to Perfection." In *American Educator* Spring, 1991.

Taba, H. *Curriculum Development: Theory and Practice*. New York: Harcourt, Brace, and Wood, 1992.

Tuckman, Bruce. *Conducting Educational Research*. New York: Harcourt, Brace, and Wood, 1992.

Tyler, R. *Basic Principles of Curriculum for Instruction*. Chicago: The University of Chicago Press, 1993.

United States Department of Education. *The Condition of Education.* Washington, DC: Government Printing Office, 2000.

Urban, Wayne and Jennings Wagoner. *American Education: A History.* New York: McGraw Hill, 2000.

Wesiman, D. *Research Strategies for Education.* Belmont, California: Wadsworth Publishing Company, 1999.

Whitehead, A. *The Aims of Education.* London: Ernest Benn Limited, 1962.

Wilder, Raymond. *Mathematics as a Cultural System.* London: Pergman, 1987.

Index